Lecture Notes in Computer Science　　10648

Commenced Publication in 1973
Founding and Former Series Editors:
Gerhard Goos, Juris Hartmanis, and Jan van Leeuwen

Editorial Board

More information about this series at http://www.springer.com/series/7409

Won-Kyung Sung · Hanmin Jung
Shuo Xu · Krisana Chinnasarn
Kazutoshi Sumiya · Jeonghoon Lee
Zhicheng Dou · Grace Hui Yang
Young-Guk Ha · Seungbock Lee (Eds.)

Information Retrieval Technology

13th Asia Information Retrieval Societies Conference, AIRS 2017
Jeju Island, South Korea, November 22–24, 2017
Proceedings

Springer

Editors

Won-Kyung Sung
Korea Institute of Science
  and Technology Information
Daejeon
Korea (Republic of)

Hanmin Jung
Korea Institute of Science
  and Technology Information
Daejeon
Korea (Republic of)

Shuo Xu
Beijing University of Technology
Beijing
China

Krisana Chinnasarn
Burapha University
Chonburi
Thailand

Kazutoshi Sumiya
Kwansei Gakuin University
Himeji, Hyogo
Japan

Jeonghoon Lee
Korea Institute of Science
  and Technology Information
Daejeon
Korea (Republic of)

Zhicheng Dou
Renmin University of China
Beijing
China

Grace Hui Yang
Georgetown University
Washington, DC
USA

Young-Guk Ha
Konkuk University
Seoul
Korea (Republic of)

Seungbock Lee
Korea Institute of Science
  and Technology Information
Daejeon
Korea (Republic of)

ISSN 0302-9743   ISSN 1611-3349   (electronic)
Lecture Notes in Computer Science
ISBN 978-3-319-70144-8   ISBN 978-3-319-70145-5   (eBook)
https://doi.org/10.1007/978-3-319-70145-5

Library of Congress Control Number: 2017957843

LNCS Sublibrary: SL3 – Information Systems and Applications, incl. Internet/Web, and HCI

Printed on acid-free paper

This Springer imprint is published by Springer Nature
The registered company is Springer International Publishing AG
The registered company address is: Gewerbestrasse 11, 6330 Cham, Switzerland

Preface

The 2017 Asian Information Retrieval Societies Conference (AIRS 2017) was the 13th instalment of the conference series, initiated from the Information Retrieval with Asian Languages (IRAL) workshop series back in 1996 in Korea. The conference was held in November 2017, at Lotte Hotel Jeju, Jeju, Republic of Korea.

The annual AIRS conference is the main information retrieval forum for the Asia Pacific region and aims to bring together academic and industry researchers, along with developers, interested in sharing new ideas and the latest achievements in the broad area of information retrieval. AIRS 2017 enjoyed contributions spanning the theory and application of information retrieval, both in text and multimedia.

All submissions were peer reviewed in a double-blind process by at least three international experts and one session chair. The final program of AIRS 2017 featured 17 full papers divided into five tracks: "IR Infrastructure and Systems," "IR Models and Theories," "Personalization and Recommendation," "Data Mining for IR," and "IR Evaluation." The program also featured short or demonstration papers.

AIRS 2017 featured two keynote speeches from Lock Pin Chew (National Security Coordination Secretariat, Singapore) and Justin Zobel (The University of Melbourne).

The conference and program chairs of AIRS 2017 extend our sincere gratitude to all authors and contributors to this year's conference. We are also grateful to the Program Committee for the great reviewing effort that guaranteed AIRS 2017 could feature a quality program of original and innovative research in information retrieval.

November 2017

Won-Kyung Sung
Hanmin Jung
Shuo Xu
Krisana Chinnasarn
Kazutoshi Sumiya
Jeonghoon Lee
Zhicheng Dou
Grace Hui Yang
Young-Guk Ha
Seungbock Lee

Organization

General Chair

Won-Kyung Sung   Korea Institute of Science and Technology Information, Republic of Korea

Organizing Chairs

Hanmin Jung   Korea Institute of Science and Technology Information, Republic of Korea
Shuo Xu   Beijing University of Technology, China
Krisana Chinnasarn   Burapha University, Thailand
Kazutoshi Sumiya   Kwansei Gakuin University, Japan

Local Organizing Chair

Jeonghoon Lee   Korea Institute of Science and Technology Information, Republic of Korea

Program Chairs

Zhicheng Dou   Renmin University of China, China
Hui Yang   Georgetown University, USA

Publicity Chair

Young-guk Ha   Konkuk University, Republic of Korea

Web and Information Chair

Seungbock Lee   Korea Institute of Science and Technology Information, Republic of Korea

Program Committee

Zhumin Chen   Shandong University, China
Ronan Cummins   University of Cambridge, UK
Hongbo Deng   Google, USA
Zhicheng Dou   Renmin University of China, China
Koji Eguchi   Kobe University, Japan
Hui Fang   University of Delaware, USA
Yi Fang   Santa Clara University, USA

Ingmar Weber	Qatar Computing Research Institute, Qatar
Hao Wu	University of Delaware, USA
Yao Wu	Twitter, Inc., USA
Takehiro Yamamoto	Kyoto University, Japan
Hui Yang	Georgetown University, USA
Yuya Yoshikawa	Chiba Institute of Technology, Japan
Hai-Tao Yu	University of Tsukuba, Japan
Mo Yu	IBM Watson, USA
Aston Zhang	University of Illinois at Urbana-Champaign, USA
Jiawei Zhang	University of Illinois at Chicago, USA
Jin Zhang	Chinese Academy of Science, China
Min Zhang	Tsinghua University, China
Mingyang Zhang	Google Inc., USA
Qi Zhang	Fudan University, China
Xin Zhao	Renmin University of China, China
Guido Zuccon	Queensland University of Technology, Australia

Steering Committee

Hsin-Hsi Chen	National Taiwan University, Taiwan, China
Zhicheng Dou	Renmin University of China, China
Wai Lam	The Chinese University of Hong Kong, Hong Kong, SAR China
Alistair Moffat	University of Melbourne, Australia
Hwee Tou Ng	National University of Singapore, Singapore
Dawei Song	Tianjin University, China
Masaharu Yoshioka	Hokkaido University, Japan

Sponsorship Committee

HakLae Kim	Korea Institute of Science and Technology Information, Republic of Korea

Contents

IR Infrastructure and Systems

IR Models and Theories

Personalization and Recommendation

IR Infrastructure and Systems

Improving Retrieval Effectiveness for Temporal-Constrained Top-K Query Processing

Hao Wu, Kuang Lu[(✉)], Xiaoming Li, and Hui Fang

University of Delaware, Newark, DE, USA
{haow,lukuang,xli,hfang}@udel.edu

Abstract. Large-scale Information Retrieval systems constantly need to strike a balance between effectiveness and efficiency. More effective methods often require longer query processing time. But if it takes too long to process a query, users would become dissatisfied and query load across servers might become unbalanced. Thus, it would be interesting to study how to process queries under temporal constraints so that search results for all queries can be returned within a specified time limit without significant effectiveness degradations. In this paper, we focus on top-K query processing for temporally constrained retrieval. The goal is to figure out what kind of query processing techniques should be used to meet the constraint on query processing time while minimizing the effectiveness loss of the search results. Specifically, we propose three temporal constrained top-K query processing techniques and then empirically evaluate them over TREC collections. Results show that all of the proposed techniques can meet the temporal constraints, and the document prioritization technique can return more effective search results.

1 Introduction

Large-scale Information Retrieval (IR) systems need to retrieve relevant search results to satisfy users' information needs, and they need to process queries efficiently. It is well known that query latency can impact users search satisfaction and search engine revenues [2,4,16]. Thus, it is important to reduce query processing time to improve search efficiency.

Modern Web search engines rely on a multi-stage query processing pipeline [10, 22,24]. It starts with retrieving top-K ranked documents on a single node using a simple retrieval function. After that, top-K results from multiple nodes are merged and re-ranked using more complicated ranking mechanisms. It is clear that top-K query processing is one of the key components in the pipeline. To improve the efficiency of top-K query processing, various dynamic pruning techniques [3,5–7,15,19,23] have been proposed to reduce the number of documents that need to be evaluated. Some methods are rank-safe for top-K results [5–7,15,19], but others can further improve the efficiency with possible degradations in the search effectiveness [3,22,27].

© Springer International Publishing AG 2017
W.-K. Sung et al. (Eds.): AIRS 2017, LNCS 10648, pp. 3–15, 2017.
https://doi.org/10.1007/978-3-319-70145-5_1

Although these efforts can improve the *average* search efficiency, the variance of query processing time is often high and the processing time of certain queries could still be longer than desired. When a query's processing time is too long, search users could potentially lose patience and discontinue the present search process as their attention was interrupted by the long response time [12,13,18]. In particular, mobile search users are often more impatient and need to get the search results in a much shorter time [17]. Commercial IR systems often set a maximum response time for query processing to avoid letting search users wait for a long time. When the query processing time is longer than the maximum allowed response time, the system will terminate the query processing and return a degraded search results. After all, it is often more desirable to return some results than no results, even if the results are not the best. Setting a maximum response time for each query is also beneficial for system load management. It can reduce the variance of query execution time, help with load balancing, and make it possible to adjust the system load when the query volume is high. Clearly, it would be necessary to study *temporally constrained query processing*, i.e., how to ensure *all* queries can be processed within a specified time limit.

The simplest temporal-constrained query processing method is forced termination. It means that, when the query processing time of a query is about to exceed the specified time limit, the system will immediately terminate the query processing and return the results. This solution can clearly meet the temporal constraint, but could significantly hurt the effectiveness because all the relevant documents at the tail of the posting lists are lost.

This paper focuses on the problem of temporally constrained top-K query processing. The goal is to study how to process a query so that it can meet a specified temporal constraint while maintaining reasonable effectiveness of the search results. We propose three methods that can potentially process queries to meet the temporal constraints without sacrificing too much on the effectiveness: (1) *rate adjustment:* which automatically adjusts the cut-off threshold rate used in the dynamic pruning methods to control the aggressiveness of the pruning; (2) *query simplification:* which simplifies long queries by removing less useful query terms; (3) *document prioritization:* which prioritizes documents and evaluates those more promising documents first.

These proposed temporally constrained top-K query processing methods are then compared and analyzed based on whether they can satisfy a strict temporal constraint for query processing time and how effective their results are compared with the results generated using rank-safe top-K query processing. Experiments are also conducted over standard TREC collections. Empirical results are consistent with our analysis. All the proposed methods are able to meet strict temporal constraints. Among them, document prioritization can generate more effective search results.

2   Related Work

Search user satisfaction is closely related to the efficiency of an IR system, and many studies proposed various techniques to reduce the query processing time.

An exhaustive query processing method evaluates every document in the collection, and then returns the top-K ranked documents. To reduce the top-K query processing time, dynamic pruning methods have been proposed [3,5–7,15,19,23,27]. Its main idea is to reduce the number of evaluated documents by skipping the documents/postings that are unlikely to be among the top-K results. Recently, Tornellotto et al. [22] proposed a selective pruning strategy that predicts the query performance and then adjusts the pruning parameter accordingly. Although this strategy can reduce the query processing time for long queries, it can not guarantee that every query can be finished within a specified time limit.

Wang et al. [25] studied the problem of temporally constrained ranking functions. They focused on identifying ranking functions that can incorporate efficiency constraints, while our focus is about query processing techniques for a given ranking function. Thus, these two studies are complementary. Lin and Trotman [9] studied a similar problem for SAAT query processing strategy.

To our best knowledge, our paper is the first work to study how to process queries using DAAT (a more commonly used query processing technique in industry) so that every query can be processed within a specified time limit without significant degradation of effectiveness.

3   Problem Formulation

Given a query, an IR system returns a list of documents that are ranked based on their relevance scores. Specifically, its *ranking function* determines how to calculate the relevance score of a document for the query based on various statistics, while its *query processing* technique determines how to traverse the inverted indexes to access the statistics and compute the scores. Both ranking functions and query processing methods can affect the effectiveness and efficiency of the IR systems. In this paper, we assume that the ranking function is fixed (e.g., Okapi BM25 in this paper), and focus on identifying useful query processing techniques under the temporal constraints. We use WAND [3] as the baseline top-K query processing technique in this paper since it is one of the commonly used DAAT query processing methods.

Not all queries have the same response time. Given a specified time limit, some queries can easily meet the temporal constraint while others cannot. Since more efficient query processing methods often hurt the effectiveness, we only need to apply them to the queries that can not meet the temporal constraint. If we have an oracle that can know the processing time for each query in advance, we could separate them into two groups based on whether they can meet the constraint. If a query can meet the temporal constraint, we can simply process them using a rank-safe strategy. Otherwise, we would need to apply a temporally constrained query processing method to meet the efficiency requirement. Although we do not have the oracle that knows the exact query time, in this paper, we leverage existing studies on performance prediction to estimate the processing time [8,11,26] and use the predicted processing time to determine whether a query needs to be processed in a different way.

4   Temporally Constrained Top-K Query Processing Methods

When processing a query, the system would need to evaluate documents, compute their relevance scores and rank them based on the scores. The processing time is closely related to the number of evaluated documents [26]. In order to meet the temporal constraint, it might be necessary to reduce the number of evaluated documents. This means that the effectiveness has to be sacrificed since relevant documents might not be evaluated in this process. Thus, a temporally constrained query processing method essentially needs to pick a subset of documents to be evaluated. The size of this subset is determined by the specified time limit, and the quality of the subset would determine the search effectiveness.

In this section, we first describe a baseline solution for temporal-constrained top-K query processing and then propose three alternative methods that can generate more effective search results while meeting the temporal constraints.

4.1   Baseline: Forced Termination

To make sure that every query's processing time meets a specified time limit, The state-of-the-art approach is to process documents using a standard top-K query processing method (such as WAND) until the processing time exceeds the limit. As soon as the processing time is about to exceed the time limit, the query processing is terminated and the incomplete search results are returned.

This early termination strategy essentially would reduce the number of evaluated documents by cutting off those at the end of posting lists. However, the posting lists are sorted based on document IDs, and the position of a document has nothing to do with the relevance score of the document. In fact, the documents with higher relevance scores are likely to be distributed throughout the posting lists. Thus, simply cutting off documents at the end of the posting lists could miss relevant documents and lead to less effective search results.

4.2   Rate Adjustment

Since the WAND method [3] makes it possible to control the number of evaluated documents through changing the cut-off threshold rate (i.e., F) [3], following the idea of previous work [22], we discuss how to leverage this mechanism to adjust the query processing time based on the specified time limit.

In WAND, the cut-off threshold rate, i.e., F, controls how aggressively the system prunes documents. When $F = 1$, the cut-off threshold equals to the score of the kth document in the ranked list, i.e., $\tau = \theta$. Therefore, the results are rank-safe, i.e., the query processing method does not affect the effectiveness of the results for the given ranking function. When $F > 1$, the cut-off is larger than the score of the kth ranked document, more documents including those relevant ones will be pruned, which means faster query processing and less effective results.

We thus propose a rate adjustment method for temporally constrained query processing. The basic idea is to first predict whether a query can be processed within the specified time limit. If so, the original rank-safe query processing method will be applied. Otherwise, F will be set to a larger value so that a more aggressive document pruning method will be used to reduce the query processing time. In this paper, we use the analytical performance modeling [26] to predict the query processing time. We now discuss two possible implementations for this method, and they are different in how to set the value of F.

The first implementation takes a binary approach, which uses a fixed cut-off threshold rate for all queries that may exceed the time limit. If the predicted time is smaller than the time limit, the query will be processed using a normal WAND algorithm, where F is set to 1. Otherwise, F will be set to a value that is larger than 1 for a more aggressive query processing strategy.

An alternative implementation to adjust the rate is to set the value based on the difference between the predicted query time and the time limit. Specifically, we have $F = \alpha * (PT - TL) + 1$, where the rate F is determined based on the different between the predicted time (PT) and the time limit (TL). When the time difference is larger, the value of F is larger and more documents will be pruned. α is a parameter that controls the relations between the time difference and the rate. A possible advantage of this implementation lies in its ability to distinguish the queries that take much longer to finish from those that are only slightly longer than the time limit.

The proposed *rate adjustment* method, no matter which implementation it uses, can certainly reduce the query processing time through more aggressive document pruning. However, it cannot guarantee that every query meets the time limit for the following two reasons. First, the efficiency prediction is not perfect. An overdue query might be falsely predicted as the ones that can be finished on time. Thus, processing this kind of queries would exceed the time limit. Second, it is difficult to choose appropriate values for F that work well for every overdue queries, because the variance of query processing time is high. If F is set to a small value, the pruning strategy might not be aggressive enough to finish processing all the queries on time, particularly for those that takes a long time to finish. But if F is set to a very large value, the system may unnecessarily prune many relevant documents, leading to worse effectiveness of the search results. To make sure that the rate-adjustment method can meet the temporal constraint, we use the hybrid version of this method, which means that we combine the rate adjustment method with the forced termination method to terminate the query processing before it is about to exceed the time limit.

4.3   Query Simplification

The processing time of a query is related to the query length (i.e., the number of unique query terms) [26]. Thus, to reduce the processing time, we could replace a long query with a shorter one that can generate similar search results.

The first step of *query simplification* is to predict the query processing time given a query. If the predicted time is longer than the specified time limit, we would select a sub-query that consists of a subset of query terms from the original query. The sub-query is selected based on two criteria: (1) its predicted processing time should be smaller than the time limit; and (2) it should generate a similar document ranking as the original query. The sub-query will then be processed using a typical top-K query processing method (e.g., WAND) and intermediate search results will be generated. After that, the scores of the intermediate results will be further adjusted using the original query terms that are not in the sub-query. Specifically, we first sort the candidate documents in the intermediate results based on their documents IDs so that it would be easier to match the posting lists for the remaining terms. After that, we go through the documents from the lowest document ID to the highest one. If a document contains a remaining query term, we will update its relevance score by adding the additional evaluation score from the remaining term. Finally, we sort the documents based on their scores from the highest to lowest, and output the results as the final results.

Since at the beginning of generating the intermediate results, the cut-off threshold grows slowly, many documents with low relevant scores are not pruned in the early phrase. As a result, the query simplification method may sometimes generate less effective results. Moreover, if a query has equally important terms, removing some terms may change the query intent and cause the system to miss highly relevant documents. Therefore, we do not expect that query simplification can output good results for every query, although it might be a good solution for some long queries. Finally, similar to the *rate adjustment* method, the query simplification method can not guarantee that every query meets the temporal constraints. Therefore, we also use a hybrid version of this method combined with the forced termination method to make sure that the query processing can be finished under the temporal constraint.

4.4   Document Prioritization

As discussed early in this section, a temporally constrained query processing method needs to identify a subset of documents that needs to be evaluated. Ideally, these documents should be those that are more likely to be relevant documents. Is there a way that we can easily identify these documents? One possible solution is to prioritize documents based on the query terms that they contain and the importance of the query terms [27]. Specifically, documents can be divided into document sets based on the query terms that they contain, and the document sets (as well as the documents in the sets) can then be prioritized based on the importance scores (such as the max scores) of the query terms that they contain. For example, given a query with two terms A and B, we can divide potentially relevant documents into three document sets: (1) documents containing both terms, (2) documents containing only term A, and (3) documents containing only term B. If A is more important than B, we assume that documents containing only term A are more important than those containing only term B. Of course, documents containing both terms would be more important than containing only one term.

Following the previous study on document prioritization [27], we propose the following method to process queries under temporal constraints. Given a query, we first divide documents into multiple document sets based on the query terms that they include. Based on their priority scores, we gradually include more document sets as the candidate documents that need to be processed until the processing time violates the temporal constraint. The system then evaluates only the documents in the selected documents sets and skips others. By applying a pivoted-based document lookup strategy similar to WAND, this method can prune documents efficiently.

No matter what the time constraint is, the document sets with the highest maximum score are always selected and evaluated. As a result, the documents with the highest scores are very likely to be processed. What is more, the document sets with low scores are likely to be discarded under tight time constraints, and the system will not spend too much time on low-score document evaluation. Finally, this method also needs to be combined with the forced termination method to ensure the satisfaction of the temporal constraint.

4.5 Discussions

As discussed earlier, the temporal-constrained query processing methods are about how to prioritize the evaluation of documents in the collection. When there is not enough time to evaluate every document, it is more desirable for the query processing method to first evaluate the documents with high relevance scores while avoiding evaluating the documents with low relevance scores.

Among all the proposed methods, the *forced termination* method is expected to generate the least effective results because its selection of documents that need to be evaluated has nothing to do with the relevance score of the documents. On the contrary, the *document prioritization* method is expected to generate the most effective results since it always evaluate the documents that are more likely to be relevant first.

5 Experiments

5.1 Experiment Design

The experiments are designed to compare the four proposed methods: (1) forced termination (CUT), (2) rate adjustment ($RATE$), (3) query simplification (QS), and (4) document prioritization (DP). Note that $RATE$ has two variants: $RATE$-B, which adjusts the rate value based on a binary decision, and $RATE$-L which adjusts the rate value linearly. We use the analytical performance modeling [26] to predict the processing time.

All the experiments are conducted over four standard TREC collections. The first three data sets were used by TREC terabyte tracks from 2004 to 2006. They are referred to as *TB04*, *TB05* and *TB06*. Each data set has 50 queries, and the document collection is the TREC gov2 collection that consists of 25.2 millions web pages crawled from the .gov domain. The fourth data set is the one used by TREC Web track from 2009 and 2012. It has 200 queries used for these four years' tracks, and the document collection is the Clueweb09 category B collection which contains 50 millions documents. This data set is denoted as *CW*.

All experiments are conducted on a single machine with dual AMD Lisbon Opteron 4122 2.2 GHz processors and 32 GB DDR3-1333 memory. WAND [3] is used as the basic top-K query processing technique, and Okapi BM25 [14] is used as the retrieval function to rank documents, and 1000 documents are returned for each query, i.e., K is set to 1000.

5.2   Performance Comparison

We now conduct experiments to compare the proposed methods under strict temporal constraints, which means all queries need to be processed within a specified time limit. As discussed in the previous section, all the proposed methods except *CUT* cannot meet the temporal constraints. One possible solution is to combine these methods with the *CUT* method. Specifically, the query processing will be terminated when the processing time is about to exceed the specified time limit. After the combination, the effectiveness may decrease because a large number of documents at the tails of posting lists are not evaluated. We compare the effectiveness of the methods (i.e., MAP@1000).

Table 1 shows the effectiveness of the proposed methods under various strict time constraints over multiple TREC terabyte collections. The parameters in the *RATE* methods (i.e., F and α) are tuned and the optimal performance for these two methods are reported in the table. It can be seen that $DP + CUT$ is the most effective method, and *CUT* is the least effective one. Moreover, although

Table 1. Performance comparison under strict temporal constraints

		15 ms	30 ms	50 ms	100 ms	200 ms	No constraint
TB04	CUT	0.100	0.151	0.200	0.240	0.246	0.250
	Rate-B + CUT	0.143	0.192	0.202	0.232	0.249	0.250
	Rate-L + CUT	0.148	0.193	0.207	0.231	0.250	0.250
	QS + CUT	0.161	0.209	0.216	0.243	0.246	0.250
	DP + CUT	**0.195**	**0.223**	**0.239**	**0.241**	0.249	0.250
TB05	CUT	0.093	0.163	0.208	0.291	0.312	0.324
	Rate-B + CUT	0.186	0.244	0.269	0.306	0.319	0.324
	Rate-L + CUT	0.188	0.247	0.273	0.308	0.318	0.324
	QS + CUT	0.196	0.252	0.273	0.307	0.319	0.324
	DP + CUT	**0.197**	**0.261**	**0.270**	**0.313**	0.319	0.324
TB06	CUT	0.086	0.165	0.225	0.270	0.283	0.285
	Rate-B + CUT	0.191	0.229	0.264	0.277	0.282	0.285
	Rate-L + CUT	0.197	0.234	0.259	0.274	0.282	0.285
	QS + CUT	0.179	0.223	0.260	0.277	0.283	0.285
	DP + CUT	**0.233**	**0.250**	**0.276**	**0.279**	0.285	0.285

the optimized parameters are used, two $RATE + CUT$ methods still do not show the best performance. Finally, $QS + CUT$ performs better than $RATE + CUT$ on the $TB04$ and $TB05$ collections but worse on the $TB06$ collection.

To further analyze the results, we report the three statistics for each method during the query processing: (1) the total number of evaluated documents; (2) the average relevance score of the evaluated documents; and (3) the progress of the processing when it is forced to terminate, which is estimated based on the number of processed postings (including those evaluated and those skipped) divided by the total number of postings of the query.

Table 2 shows the average values of these statistics when $TL = 15$ ms. Let us start with the CUT method. Recall that this method essentially uses the WAND method to process query and terminate the process of a query when it is about to exceed the time limit. We can see that this method goes slowly and only finishes processing around 20% of the postings within the time limit. Assuming relevant documents are evenly distributed across the postings, around 80% of relevant documents are not included in the search results. What is more, the average score of the evaluated documents is lower than other methods. This suggests that most of the evaluated documents have low relevance scores and are more likely to be non-relevant.

Table 2. Statistics comparison when process is forced to terminate ($TL = 15$ ms)

		# of Evals	Avg. score	Process
TB04	CUT	29,818	5.89	18%
	Rate-B + CUT	13,212	7.62	45%
	Rate-L + CUT	13,489	7.73	49%
	QS + CUT	30,349	6.59	59%
	DP + CUT	7,257	11.9	59%
TB05	CUT	33,227	5.60	18%
	Rate-B + CUT	13,313	7.31	53%
	Rate-L + CUT	13,680	7.31	55%
	QS + CUT	28,739	7.07	64%
	DP + CUT	10,374	11.0	62%
TB06	CUT	32,033	5.89	16%
	Rate-B + CUT	13,994	7.70	54%
	Rate-L + CUT	15,436	7.69	53%
	QS + CUT	33,876	6.83	68%
	DP + CUT	7,149	12.1	72%

Compared with the CUT method, two $RATE + CUT$ methods are able to make more progress in the query processing but with fewer number of evaluated documents. This indicates that more aggressive pruning methods can skip the evaluation of non-relevant documents (i.e., those with lower relevance scores), which lead to better effectiveness.

$DP + CUT$ shows even more advantages based on these statistics. First, its average scores of evaluated documents are significantly higher than other methods. It confirms our hypothesis that DP is able to avoid evaluating documents with lower relevance scores. What is more, DP can make more progress during the query processing. Although on average DP evaluates fewer documents, it can return more effective results than other methods.

$QS + CUT$ reduces the number of processed query terms in exchange for better efficiency. We can see that it usually evaluates more documents within the time limit because reducing the number of query terms can greatly reduce the time spent on evaluating a document. When comparing the average score of the evaluated documents, its value is only slightly smaller than those for the $RATE + CUT$ methods. Considering it uses fewer query terms, this number still indicates the good quality of its evaluated documents. As shown in Table 1, $QS + CUT$ performs better than $RATE + CUT$ over the $TB04$ and $TB05$ collections but worse on the $TB06$ collection. After further analysis, we find that one possible reason is that many queries in the $TB06$ data set contain phrases. For these queries, deleting query terms may change the meaning of the original queries, which would lead to worse effectiveness.

Table 3. Performance comparison under strict time constraints on the CW collection

	15 ms	30 ms	50 ms	100 ms	200 ms	No constraint
CUT	0.049	0.084	0.117	0.143	0.173	0.188
RATE-B + CUT	0.075	0.105	0.123	0.142	0.154	0.188
RATE-L + CUT	0.074	0.098	0.116	0.132	0.152	0.188
QS + CUT	0.093	0.117	0.131	0.151	0.172	0.188
DP + CUT	**0.098**	**0.130**	**0.149**	**0.166**	**0.178**	0.188

Finally, we evaluate the proposed methods on a larger collection, i.e., CW, whose document collection size is as twice large as the one used for the terabyte tracks. As more documents need to be processed, the expected average query processing time would be longer. The performance comparison under the strict time constraints are reported in Table 3. Clearly, $DP + CUP$ is more effective than other methods. Since Web search engines would need to process much larger data collections, we expect that the advantage of the proposed methods remains even when TL is set to a larger value.

5.3   More Analysis

We also conduct experiments to examine how the query length would impact the search effectiveness. The results are summarized in Table 4. It is clear that

Table 4. Effectiveness (MAP@1000) for queries with different length on TB04–06 (TL = 15 ms)

Query length	1	2	3	4	>4
CUT	0.159	0.145	0.097	0.056	0.012
RATE-B + CUT	0.159	0.254	0.158	0.156	0.027
RATE-L + CUT	0.159	0.256	0.169	0.160	0.034
QS + CUT	0.159	0.224	0.159	0.193	0.070
DP + CUT	**0.159**	**0.249**	**0.206**	**0.210**	**0.072**

Table 5. Performance of conjunctive (AND) query processing

	Processing time (ms)	MAP
TB04	52.3	0.225
TB05	31.2	0.236
TB06	34.7	0.247
CW	109.7	0.150

the *document prioritization* method consistently outperform other methods for queries with various lengths.

This paper focuses on disjunctive query processing mode (i.e., OR mode) because it can generate more effective search results. However, since the conjunctive (i.e., AND) mode [1,20,21] can greatly improve the efficiency, it would be interesting to compare it with the proposed temporally constrained methods. Table 5 summarizes the performance when queries are processed using the conjunctive mode. We report the results for both effectiveness (i.e., MAP) and efficiency (i.e., average processing time in ms). It is clear that this method cannot meet the time constraint when TL is set to a small value such as 15 ms. Moreover, its effectiveness is worse than some of the proposed methods such as *DP*. For example, for the *TB04* data set, the processing time of the AND mode is 52.3 ms, and its MAP is 0.225, which is worse than the effectiveness of *DP* (i.e., 0.239) when $TL = 50$ ms.

6   Conclusions and Future Work

Although there are a lot of previous studies about efficiency of IR systems, none of them guarantees that all queries can be processed within a specified time limit. To our best knowledge, this paper is the first one that explores the problem of temporally constrained top-K query processing for DAAT, i.e., how to dynamically adjust query processing strategy to meet time constraints and optimize effectiveness under the temporal constraints. In this paper, we propose three temporal-constrained query processing methods and then evaluate them on TREC collections. Empirical results are consistent with our analysis and indicate that the proposed document prioritization methods can generate the most effective results while meeting temporal constraints.

There are many interesting directions for the future work. First, we plan to explore more methods to process query in a temporally constrained environment. Second, we focus on only the DAAT methods in this paper and plan to study other index traversing method such as TAAT. Finally, it would be interesting to study how to apply these techniques in the distributed environment.

Acknowledgements. This research was supported by the U.S. National Science Foundation under IIS-1423002.

References

1. Asadi, N., Lin, J.: Effectiveness/efficiency tradeoffs for candidate generation in multi-stage retrieval architectures. In: Proceedings of SIGIR 2013 (2013)
2. Barroso, L.A., Dean, J., Hölzle, U.: Web search for a planet: the Google cluster architecture. IEEE Micro **23**(2), 22–28 (2003)
3. Broder, A.Z., Carmel, D., Herscovici, M., Soffer, A., Zien, J.: Efficient query evaluation using a two-level retrieval process. In: Proceedings of CIKM 2003 (2003)
4. Brutlag, J.D., Hutchinson, H., Stone, M.: User preference and search engine latency. In: Proceedings of JSM (2008)
5. Chakrabarti, K., Chaudhuri, S., Ganti, V.: Interval-based pruning for top-k processing over compressed lists. In: Proceedings of ICDE 2011 (2011)
6. Dimopoulos, C., Nepomnyachiy, S., Suel, T.: A candidate filtering mechanism for fast top-k query processing on modern CPUs. In: Proceedings of SIGIR 2013 (2013)
7. Ding, S., Suel, T.: Faster top-k document retrieval using block-max indexes. In: Proceedings of SIGIR 2011 (2011)
8. Jeon, M., Kim, S., Hwang, S., He, Y., Elnikety, S., Cox, A.L., Rixner, S.: Predictive parallelization: taming tail latencies in web search. In: Proceedings of SIGIR 2014 (2014)
9. Lin, J., Trotman, A.: Anytime ranking for impact-ordered indexes. In: Proceedings of the ICTIR (2015)
10. Liu, T.-Y.: Learning to rank for information retrieval. Found. Trends Inf. Retr. **3**(3), 225–331 (2009)
11. Macdonald, C., Tonellotto, N., Ounis, I.: Learning to predict response times for online query scheduling. In: Proceedings of SIGIR 2012 (2012)
12. Miller, R.B.: Reponse time in man-computer conversational transactions. In: Proceedings of the AFIPS, pp. 267–277 (1968)
13. Neilsen, J.: Usability Engineering. Elsevier, New York (1994)
14. Robertson, S.E., Walker, S., Jones, S., Hancock-Beaulieu, M.M., Gatford, M.: Okapi at TREC-3. In: Proceedings of TREC-3 (1995)
15. Rossi, C., de Moura, E.S., Carvalho, A.L., da Silva, A.S.: Fast document-at-a-time query processing using two-tier indexes. In: Proceedings of SIGIR 2013 (2013)
16. Schurman, E., Brutlag, J.: Performance related changes and their user impact. In: Velocity - Web Performance and Operations Conference (2009)
17. Shmueli-Scheuer, M., Li, C., Mass, Y., Roitman, H., Schenkel, R., Weikum, G.: Best-effort top-k query processing under budgetary constraints. In: Proceedings of ICDE, pp. 928–939 (2009)

18. Shneiderman, B.: Reponse time and display rate in human performance with computers. ACM Comput. Surv. **16**(3), 265–285 (1984)
19. Strohman, T., Croft, B.W.: Efficient document retrieval in main memory. In: Proceedings of SIGIR 2007 (2007)
20. Takuma, D., Yanagisawa, H.: Faster upper bounding of intersection sizes. In: Proceedings of SIGIR 2013 (2013)
21. Tatikonda, S., Cambazoglu, B.B., Junqueira, F.P.: Posting list intersection on multicore architectures. In: Proceedings of SIGIR 2011 (2011)
22. Tonellotto, N., Macdonald, C., Ounis, I.: Efficient and effective retrieval using selective pruning. In: Proceedings of WSDM 2013 (2013)
23. Turtle, H., Flood, J.: Query evaluation: strategies and optimizations. Inf. Process. Manage. **31**(6), 831–850 (1995)
24. Wang, L., Lin, J., Metzler, D.: Learning to efficiently rank. In: Proceedings of SIGIR 2010 (2010)
25. Wang, L., Metzler, D., Lin, J.: Ranking under temporal constraints. In: Proceedings of the CIKM, pp. 79–88 (2010)
26. Wu, H., Fang, H.: Analytical performance modeling for top-k query processing. In: Proceedings of CIKM 2014 (2014)
27. Wu, H., Fang, H.: Document prioritization for scalable query processing. In: Proceedings of CIKM 2014 (2014)

A Boosted Supervised Semantic Indexing for Reranking

Takuya Makino(✉) and Tomoya Iwakura

Fujitsu Laboratories Ltd., Kawasaki, Japan
{makino.takuya,iwakura.tomoya}@jp.fujitsu.com

Abstract. This paper proposes a word embedding-based reranking algorithm with a boosting. The algorithm converts queries and documents into sets of word embeddings represented by vectors and reranks documents according to a similarity defined with the word embeddings as in Latent Semantic Indexing (LSI) and Supervised Semantic Indexing (SSI). Compared with LSI and SSI, our method uses top-n irrelevant documents of a relevant document of each query for training a reranking model. Furthermore, we also propose application of a boosting to the reranking model. Our method uses the weights of training samples decided by AdaBoost as coefficients for updating model, therefore, highly weighted samples are aggressively learned. We evaluate the proposed method with datasets created from English and Japanese Wikipedia respectively. The experimental results show that our method achieves better mean average precision than LSI and SSI.

1 Introduction

Learning to rank is a machine learning-based method for ranking of information retrieval. With a given training data, learning to rank methods learn weights of features such as rankings given by search engines and content similarities between a query and a document. Documents are ranked based on the scores given by the learned model. One of the approaches of learning to rank uses hand-crafted features [8,12], such as tf-idf, the title, URL, PageRank and other information. Another approach is a way of learning a model by considering pairs of words between the two texts. The difficulty is that such feature spaces are very large. In order to deal with the difficulty, Supervised Semantic Indexing (SSI) [1] was proposed. For training a ranker, SSI uses a pairwise learning to rank approach that maps a word into a low dimensional vector, which is called as a word embedding, for calculating score. By mapping words into word embeddings, SSI can solve the feature space problem. The SSI approach is suitable for queries represented by natural language used in systems such as an FAQ search and patent searches.

However, the training of SSI does not meet the setting of reranking. The original SSI only handles a negative document for each relevant document of each query (Fig. 1a). In other words, the training setting is different from reranking of top-n search engine results.

W.-K. Sung et al. (Eds.): AIRS 2017, LNCS 10648, pp. 16–28, 2017.
https://doi.org/10.1007/978-3-319-70145-5_2

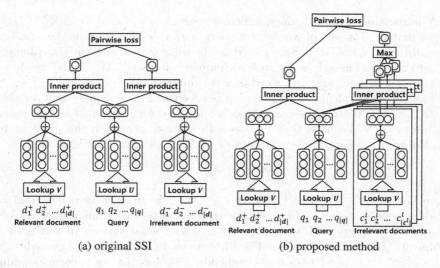

(a) original SSI   (b) proposed method

Fig. 1. Overviews of a training of original SSI and our reranking model. $\oplus$ is an element wise addition of vectors. $\mathbf{V}$ is common for relevant and irrelevant documents.

In order to improve the SSI for reranking, we propose the following:

- A training of SSI with irrelevant documents for each relevant document of each query. In our setting, a relevant document and its irrelevant documents are used as a training sample and the training aims at learning a model that gives the highest score to relevant document of each training sample (Fig. 1b).
- Application of a boosting: In addition, in order to get further improved ranking accuracy, we also propose an application of an AdaBoost, which is a boosting algorithm, to the proposed SSI for reranking. Our method uses the weights of training samples decided by the AdaBoost as coefficients for updating model. Therefore, training samples with high weights given by the AdaBoost are aggressively learned.

We first introduce SSI in Sect. 2, then, we propose Reranking SSI with a boosting in Sect. 3. We report the evaluation results of the proposed method with data sets that are created from Japanese and English Wikipedia in Sect. 4. The experimental results show that our method achieves better mean average precision than SSI.

2   Supervised Semantic Indexing

2.1   Preliminary

Equation (1) is a naive score function of SSI, which calculates a similarity between a query $\mathbf{q} \in \mathbb{R}^{N \times 1}$ and a document $\mathbf{d} \in \mathbb{R}^{N \times 1}$.

$$f(\mathbf{q}, \mathbf{d}) = \mathbf{q}^\top \mathbf{W} \mathbf{d}, \tag{1}$$

N is the size of a vocabulary, which is a set of words, and $\mathbf{W} \in \mathbb{R}^{N \times N}$ is a weight matrix of a pair of words in a query and a document. q_i is the value of i-th word in a vocabulary for a query. d_j is the value of j-th word in a vocabulary for a document. The values are such as frequency and tfidf. W_{ij} is the weight of the pair of i-th word and j-th word in a vocabulary.

This function calculates relationship between words in a query and a document. However, there are some problems such as data sparseness and memory requirements ($\mathbf{W}$ requires $O(N^2)$ space). In order to deal with the space problem, Bai et al. [1] proposed to decompose $\mathbf{W}$ into low dimensional matrices.

$$\mathbf{W} = \mathbf{U}^\top \mathbf{V} + \mathbf{I}, \tag{2}$$

where $\mathbf{U} \in \mathbb{R}^{K \times N}$, $\mathbf{V} \in \mathbb{R}^{K \times N}$ and $\mathbf{I} \in \mathbb{R}^{N \times N}$. $\mathbf{I}$ is an identity matrix and K is a dimension of a vector. In other words, $\mathbf{U}$ consists of K-dimensional vectors of words in a query and $\mathbf{V}$ consists of K-dimensional vectors of words in a document. $\mathbf{U}$ and $\mathbf{V}$ can be viewed as a matrix that consists of word embeddings and a word is mapped to a word embedding by looking up a corresponding column vector of those matrices. Equation (1) can be rewritten as follows:

$$f(\mathbf{q}, \mathbf{d}) = \mathbf{q}^\top (\mathbf{U}^\top \mathbf{V} + \mathbf{I}) \mathbf{d},$$
$$= (\sum_i^N q_i \mathbf{u}_i) \cdot (\sum_i^N d_i \mathbf{v}_i) + \sum_i^N q_i d_i,$$

where $\mathbf{u}_i$ is the i-th column vector of $\mathbf{U}$ and $\mathbf{v}_i$ is the i-th column vector of $\mathbf{V}$. SSI calculates a score that is a summation of two terms. The first term $(\sum_i^N q_i \mathbf{u}_i) \cdot (\sum_i^N d_i \mathbf{v}_i)$ is the inner product between the vector of a query and the vector of a document in a low dimensional space. The vectors of a query and a document are the summations of word embeddings of a query and a document weighted by the value of a corresponding word such as tf-idf. The second term $\sum_i^N q_i d_i$ is the surface similarity, which is the inner product between $\mathbf{q}$ and $\mathbf{d}$.

For a given query, documents are sorted in a descending order according to scores that are calculated by a model.

2.2 A Training Method

Let $\mathcal{R} = \{(\mathbf{q}_i, \mathbf{d}_i^+, \mathbf{d}_i^-)\}_{i=1}^m$ be a set of training samples where i-th training sample is a tuple of query $\mathbf{q}_i$, a relevant document $\mathbf{d}_i^+$ and an irrelevant document $\mathbf{d}_i^-$. The purpose of training SSI is deriving $\mathbf{U}$ and $\mathbf{V}$ that minimizes the following on $\mathcal{R}$:

$$\sum_{(\mathbf{q}_i, \mathbf{d}_i^+, \mathbf{d}_i^-) \in \mathcal{R}} \max(0, 1 - f(\mathbf{q}_i, \mathbf{d}_i^+) + f(\mathbf{q}_i, \mathbf{d}_i^-)).$$

Algorithm 1 shows a pseudo code for the training algorithm. At first, $\mathbf{U}$ and $\mathbf{V}$ are initialized by normal distribution with mean zero and standard deviation one.

#Training data: $\mathcal{R} = \{(\mathbf{q}_i, \mathbf{d}_i^+, \mathbf{d}_i^-)\}_{i=1}^m$
#The learning rate: λ
#The maximum iteration of SSI: P
Initialize: $\mathbf{U}, \mathbf{V}$
$p = 1$ **while** $p \leq P$ **do**
    for $i = 1...m$ **do**
        if $1 - f(\mathbf{q}_i, \mathbf{d}_i^+) + f(\mathbf{q}_i, \mathbf{d}_i^-) > 0$ **then**
            Update $\mathbf{U}$ by Eq. (3)
            Update $\mathbf{V}$ by Eq. (4)
        end
    end
    $p++$
end
Return $\mathbf{U}, \mathbf{V}$

Algorithm 1. A Supervised Semantic Indexing [1].

Then SSI picks up a training sample from $\mathcal{R}$. If SSI gives a higher score to the pair of a query and an irrelevant document than the pair of the query and a relevant document, $\mathbf{U}$ and $\mathbf{V}$ are updated with Eqs. (3) and (4):

$$\mathbf{U} = \mathbf{U} + \lambda \mathbf{V}(\mathbf{d}^+ - \mathbf{d}^-)\mathbf{q}^\top, \tag{3}$$
$$\mathbf{V} = \mathbf{V} + \lambda \mathbf{U}(\mathbf{d}^+ - \mathbf{d}^-)\mathbf{q}^\top, \tag{4}$$

where λ is a learning ratio.

3   Boosted Supervsed Semantic Indexing for Reranking

In this section, we describe the training algorithm of SSI for reranking (ReSSI). Then, we describe an application of a boosting to ReSSI (Boosted ReSSI).

3.1   Extension of SSI to Reranking

Algorithm 2 shows a pseudo code for the training algorithm of ReSSI and Fig. 1 shows an overview of ReSSI. Let $\mathcal{R}'$ be a set of training samples and each sample is a tuple of query $\mathbf{q}$, a relevant document $\mathbf{d}^+$ and the set of irrelevant documents $\{\mathbf{c}^l\}_{l=1}^L$ where $\mathbf{c}^l \in \mathbb{R}^{N \times 1}$.

ReSSI learns a model to rank the relevant document $\mathbf{d}^+$ of each query $\mathbf{q}$ higher than its corresponding irrelevant documents $\{\mathbf{c}^l\}_{l=1}^L$.

In order to use L documents as negative samples for a query, we propose a training method for reranking. ReSSI uses the irrelevant document $\mathbf{c}_i^j$ that has the highest score $f(\mathbf{q}_i, \mathbf{c}_i^l)$ as $\mathbf{c}_i^-$ in the irrelevant documents $\{\mathbf{c}_i^l\}_{l=1}^L$ of query $\mathbf{q}_i$. If $(1 - f(\mathbf{q}_i, \mathbf{d}^+) + f(\mathbf{q}_i, \mathbf{c}_i^-) > 0)$ is satisfied, a current model is updated by Eqs. (5) and (6):

$$\mathbf{U} = \mathbf{U} + \lambda \epsilon_i \mathbf{V}(\mathbf{d}_i^+ - \mathbf{c}_i^-)\mathbf{q}_i^\top, \tag{5}$$
$$\mathbf{V} = \mathbf{V} + \lambda \epsilon_i \mathbf{U}(\mathbf{d}_i^+ - \mathbf{c}_i^-)\mathbf{q}_i^\top, \tag{6}$$

#Training data: $\mathcal{R}' = \{(\mathbf{q}_i, \mathbf{d}_i^+, \{\mathbf{c}_i^l\}_{l=1}^L)\}_{i=1}^m$
#The maximum iteration of SSI: P
#The weights of samples: $\{\epsilon_i\}_{i=1}^m$
#The learning ratio: λ
$\mathcal{R}' = \{(\mathbf{q}_i, \mathbf{d}_i^+, \{\mathbf{c}_i^l\}_{l=1}^L)\}_{i=1}^m$
Initialize: $\mathbf{U}, \mathbf{V}$
$p = 1$
while $p \le P$ **do**

    for $i = 1...m$ **do**

        $\mathbf{c}_i^- = \arg\max_{\mathbf{c}_i^l} f(\mathbf{q}_i, \mathbf{c}_i^l)$ **if** $1 - f(\mathbf{q}_i, \mathbf{d}_i^+) + f(\mathbf{q}_i, \mathbf{c}_i^-) > 0$ **then**

            Update $\mathbf{U}$ by Eq. (5)

            Update $\mathbf{V}$ by Eq. (6)

        end

    end

    $p + +$

end
Return $\mathbf{U}, \mathbf{V}$

Algorithm 2. A Supervised Semantic Indexing for reranking: ReSSI $(\mathcal{R}', P, \{\epsilon_i\}_{i=1}^m, \lambda)$

where $\{\epsilon\}_{i=1}^m$ are the weights of training samples. We set $\epsilon_i = 1$ $(1 \le i \le m)$ for ReSSI without boosting. If we use $L = 1$ and $\epsilon_i = 1$ $(1 \le i \le m)$, this is equivalent to the original SSI.

3.2 Application of Boosting

We apply a variant of AdaBoost [9] to ReSSI. The original AdaBoost was designed for the classification problem. Here, we propose to apply the method designed for structured prediction [11] to a learning to rank.

We show a pseudo code of the application of the boosting to ReSSI in Algorithm 3. In the first iteration, ReSSI is trained with the initial weights of samples $\epsilon_i = 1/m (1 \le i \le m)$.

Then, the boosting learner updates the weights of training samples. The boosting learner assigns larger weights to training samples that are incorrectly ranked. To realize this, we define a loss for a training sample $(\mathbf{q}_i, \mathbf{d}_i^+, \{\mathbf{c}_i^l\}_{l=1}^L)$ as follows:

$$s_t(\mathbf{q}_i, \mathbf{d}_i^+, \{\mathbf{c}_i^l\}_{l=1}^L) = f_t(\mathbf{q}_i, \mathbf{d}_i^+) - f_t(\mathbf{q}_i, \mathbf{c}_i^-),$$

where, $\mathbf{c}_i^- = \arg\max_{\mathbf{c}_i^l} f_t(\mathbf{q}_i, \mathbf{c}_i^l)$ and $f_t(\mathbf{q}_i, \mathbf{c}_i^l) = \mathbf{q}_i^\top (\mathbf{U}_t^\top \mathbf{V}_t + \mathbf{I}) \mathbf{c}_i^l$.

After the boosting learner learns a weak learner at time t, it searches a confidence-value α_t that satisfies $\tilde{Z}_t(\tilde{\alpha}_t) < 1$:

$$\tilde{Z}_t(\tilde{\alpha}_t) = \sum_{i=1}^m w_{t,i} e^{-\tilde{\alpha}_t s_t(\mathbf{q}_i, \mathbf{d}_i^+, \{\mathbf{c}_i^l\}_{l=1}^L)},$$

#Training data: $\mathcal{R}' = \{(\mathbf{q}_i, \mathbf{d}_i^+, \{\mathbf{c}_i^l\}_{l=1}^L)\}_{i=1}^m$
#The iteration of SSI training: P
#The learning ratio of SSI training: λ
Initialize: $U = \{\}, V = \{\}, A = \{\}$
$t = 1$
set initial value: $w_{1,i} = \frac{1}{m}$ (for $1 \leq i \leq m$)
while $t \leq T$ **do**
 | $\mathbf{U}_t, \mathbf{V}_t = \text{ReSSI}(\mathcal{R}', P, \{w_{t,i}\}_{i=1}^m, \lambda)$
 | Find $\tilde{\alpha}_t$ that satisfies $\tilde{Z}_t(\tilde{\alpha}_t) < 1$.
 | $U \leftarrow U \cup \{\mathbf{U}_t\}$
 | $V \leftarrow V \cup \{\mathbf{V}_t\}$
 | $A \leftarrow A \cup \{\tilde{\alpha}_t\}$
 | **for** $i = 1...m$ **do**
 | | Update sample weights by Eq. (7)
 | **end**
 | $t++$
end
Return U, V, A

Algorithm 3. A Boosted Supervised Semantic Indexing

where $w_{t,i}$ is the weight of i-th sample at time t and e is Napier's constant. We find $\tilde{\alpha}_t$ with the same method used in [11].

After obtaining α_t, the weight of each sample is updated as follows:

$$w_{t+1,i} = w_{t,i} \frac{e^{-\alpha_t s_t(\mathbf{q}_i, \mathbf{d}_i^+, \{\mathbf{c}_i^l\}_{l=1}^L)}}{\tilde{Z}_t(\alpha_t)}. \tag{7}$$

After training a boosting learner, we obtain t ReSSI models and their confidence values. Given a query and a document, a final ranker calculates scores by using all ReSSI models. The score of a document given by a final ranker is a summation of scores given by ReSSI models.

$$f^*(\mathbf{q}, \mathbf{d}) = \sum_{t=1}^T \tilde{\alpha}_t f_t(\mathbf{q}, \mathbf{d}).$$

4 Experiments

4.1 Dataset

We conduct our experiments on datasets generated from English Wikipedia and Japanese Wikipedia as in [1] with the following steps[1].

---

[1] We used https://dumps.wikimedia.org/jawiki/20160407/jawiki-20160407-pages-art icles.xml.bz2 and https://dumps.wikimedia.org/enwiki/20160901/enwiki-20160901-pages-articles.xml.bz2. Retrieved October 14, 2016.

- (d1) Preprocessing: Before generating data sets, we removed markup from articles in Wikipedia by using regular expressions. For extracting words from English articles, we used white space. For tokenizing Japanese articles, we used MeCab[2], which is a Japanese morphological analyzer.
- (d2) Selecting Wikipedia articles as queries: 10,000 articles for training and 1,000 articles for test are randomly sampled from each Wikipedia archive as queries.
- (d3) Collecting positive samples of queries: We used Wikipedia articles linked by each query Wikipedia article collected at (d2) as positive samples of the query.
- (d4) Collecting negative samples of queries: We collected negative samples by searching articles in each Wikipedia archive with words in each query article. We used a full text search engine Elasticsearch[3]. OR search of words in each query article was used for collecting irrelevant articles as negative samples. We collected $L = 10$ articles other than positive documents for each query. In the training phase, for generating training data for a query that has multiple relevant documents, we assign the same irrelevant documents obtained with the query to each relevant document. For evaluation, a set of the relevant documents and the L irrelevant documents of each query is given to a ranking algorithm.

Table 1. Size of training and test data sets.

English		Japanese	
Train	Test	Train	Test
72,816	7,820	243,600	25,358

Table 1 lists the size of training data. Average numbers of words that are in vocabulary of training data and test data are 241 and 465 respectively.

4.2   Methods to Be Compared

The following algorithms are compared with our methods; ReSSI and Boosted ReSSI.

tfidf. Documents are sorted in descending order based on an inner product between a vector of query and a vector of a document. The weight of each word in a query and a document is normalized tfidf. A normalized tfidf is an original tfidf divided by the number of words in the query and the document respectively.

---

[2] https://github.com/taku910/mecab. Retrieved October 14, 2016.
[3] https://www.elastic.co/downloads/past-releases/elasticsearch-1-7-1,   Retrieved October 14, 2016.

Table 2. Evaluation results on enwiki and jawiki. The scores with $*$ significantly differs with ones of Boosted ReSSI with $p \leq 0.01$. The scores with $\dagger$ significantly differs with ones of ReSSI with $p \leq 0.01$.

	MAP	
	enwiki	jawiki
tfidf	$0.016^{*\dagger}$	$0.011^{*\dagger}$
LSI	$0.314^{*\dagger}$	$0.069^{*\dagger}$
SSI	$0.573^{*\dagger}$	$0.392^{*\dagger}$
ReSSI	0.623^*	0.482^*
Boosted ReSSI	$\mathbf{0.634^{\dagger}}$	$\mathbf{0.497^{\dagger}}$

LSI. Latent Semantic Indexing [6]. LSI is an unsupervised dimensional reducing model. We used SVDLIBC[4],[5] for obtaining latent parameters. Documents are sorted in descending order based on inner product with the vector of a query in a latent space.

SSI Algorithm 1. Original paper randomly selected a negative sample of each relevant document. In this paper, we select a negative sample from the document that has the highest score in a search result obtained with a query of each relevant document.

ReSSI. This is our proposed reranking-based SSI. The negative sample of each query is selected from the top 10 search result of a search engine obtained with the query.

Boosted ReSSI. This is the boosting version of ReSSI.

We set the size of the dictionary to $N = 30,000$, and the maximum iteration number for SSI and ReSSI to $P = 30$. For the proposed method, we set the maximum iteration number for boosting to $T = 5$. Dimensions of word embeddings in SSI, ReSSI and Boosted ReSSI are set to $K = 50$. $\mathbf{U}$ and $\mathbf{V}$ are initialized by normal distribution with mean zero and standard deviation one. The learning ratios of each algorithm is following. For SSI and ReSSI, we used $\lambda = 0.01$. For Boosted ReSSI, we used $\lambda = 0.01 \times m$ this is because the larger m is given, the less weight is given for each training sample by the AdaBoost. We implemented SSI, ReSSI and Boosted ReSSI with C++.

4.3   Evaluation Metric

We use Mean Average Precision (MAP), the mean of average precision over a collection of questions, as our evaluation metric. For evaluation, a set of the r_i relevant documents and the L irrelevant documents of each query $\mathbf{q}_i$ is ranked with a reranking algorithm first. Let $[\mathbf{d}_i^1, ..., \mathbf{d}_i^{r_i+L}]$ be a list of documents of

---

[4] https://tedlab.mit.edu/~dr/SVDLIBC/. Retrieved October 14, 2016.
[5] We used following options: $-$d 50 $-$i 5 $-$e 1e$-$30 $-$a las2 $-$k 1e$-$6.

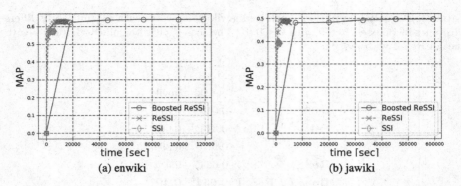

Fig. 2. Accuracy on each training time. MAPs are calculated on test data.

$\mathbf{q}_i$ ranked by a reranking algorithm. $\mathbf{d}_i^1$ is the top ranked document for $\mathbf{q}_i$. We calculate MAP as follows:

$$\text{MAP} = \frac{1}{C} \sum_{i=1}^{C} \frac{\sum_{j=1}^{r_i+L} P_{i,j}}{r_i},$$

where, C is the size of test data, $P_{i,j}$ is the precision if the j-th document of $\mathbf{q}_i$ is relevant, otherwise zero. $P_{i,j}$ is calculated as follows:

$$P_{i,j} = \frac{R_j}{j} [[\mathbf{d}_i^j \text{ is a relevant document}]],$$

where R_j is the number of relevant documents in the top-j documents and $[[\pi]]$ is 1 if a proposition π holds and 0 otherwise.

$\frac{\sum_{j=1}^{r_i+L} P_{i,j}}{r_i}$ is 1 if all r_i relevant documents of $\mathbf{q}_i$ are ranked in top-r_i for all questions, otherwise it becomes smaller than 1.

4.4   Accuracy

Table 2 shows main result of our experiments on enwiki and jawiki. We see that ReSSI shows higher accuracy than tfidf, LSI and SSI. Furthermore, Boosted ReSSI outperforms all the other methods. For significance test, we used paired t-test. For calculating p-value, we used average precision for each query. There are significant differences between ReSSI and tfidf, LSI and SSI. Furthermore, there are also significant differences between Boosted ReSSI and the others. MAP of tfidf is lower than other methods. This indicates that predicting link relation between a query and a document based on surface word matching similarity is difficult.

4.5 Transition of Accuracy by Training Time

We compared MAPs for each model with different sizes of training data. In Fig. 2, we plot MAP of test data on each epoch in the training phase. We use the seconds of epoch as horizontal axis instead of the number of epochs. For the boosting algorithm, we use the sum of the seconds by each boosting round as horizontal axis. SSI converges faster than ReSSI since SSI fixes negative examples. However, the performance of ReSSI is better than SSI.

4.6 Accuracy on Different Training Data Size

Figure 3 shows the results of different sizes of training data. For both datasets, we can see that our proposed methods show higher accuracy than SSI not only the larger training data set but also smaller training data sets.

4.7 Reranking Speed

We compared the times of reranking of ReSSI and Boosted ReSSI. For measuring times, we used models that are trained on 50% samples that are randomly sampled from enwiki and jawiki respectively. Times are measured on Intel(R) Xeon(R) CPU E5-2630 v3 @ 2.40 GHz and we used a single CPU for each model. Table 3 shows the average seconds of predicting of our proposed methods over test queries. The time of reranking of Boosted ReSSI is almost five times, which is the number of boosting round in our experiments, times larger than ReSSI.

For the AdaBoost to structured perceptron [11], merging weak learners by addition is effective. However, for Boosted ReSSI, the approach is not effective. This is because the boosting learner needs to calculate the product of matrices $\mathbf{U}_t$ and $\mathbf{V}_t$ when merging a weak learner by addition and it requires $O(N^2)$ memory complexity and $O(|\mathbf{q}||\mathbf{d}|N)$ computational complexity where $|\mathbf{q}|$ and $|\mathbf{d}|$ are the number of non zero values in $\mathbf{q}$ and $\mathbf{d}$ respectively.

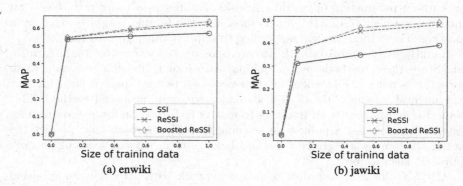

Fig. 3. Transition of Accuracy with different sizes of training data. We randomly sampled 10% and 50% of training data from enwiki and jawiki respectively.

However, if we have multiple CPU or multicores, we can easily improve the reranking speed by distributing the calculation because we can independently calculate scores given by each model created in boosting.

Table 3. Average seconds of reranking

	enwiki	jawiki
ReSSI	0.081	0.234
Boosted ReSSI	0.386	1.101

5   Related Works

For obtaining word vectors, there are ways such as matrix factorization based approach and learning to predict words given context. Latent Semantic Indexing [6], probabilistic Latent Semantic Analysis [2] and Latent Dirichlet Allocation [4] are well-known matrix factorization based vector space models without hand-crafted features. These models can map a word into a low dimensional vector for calculating a score between a query and a document. Skip-gram, CBOW and GloVe [14,16] learns to predict words given context words. For predicting a word, these methods calculate inner product between a word and its context word. In contrast, our method learns word embeddings and models to rank a relevant document higher than an irrelevant document.

Supervised learning to rank models that map a word into a low dimensional vector are proposed in various tasks such as sentiment classification and image retrieval task [3,10,15,18]. Our focus is training methods of a reranking model for the pair of a query and a document.

For updating a ranking model, Bespalov et al. [3] proposed WARP loss that is based on the rank of a relevant document. Recent works of optimization methods for learning parameters of neural networks modifies a learning ratio based on square of gradient [7,13,21]. Our method uses the sample weight decided by a boosting. These methods can be applied to improve our proposed method.

RankBoost [8] and AdaRank [19] are boosting algorithms for ranking problem. Since these methods use a ranking feature or a decision tree as a weak learner, it is not scalable when the dimension of a feature space is large like our experimental setting. AdaMF [17] is similar with our proposed method. This method is an application of boosting to matrix factorization for a recommender system. AdaMF uses a point-wise learning to rank approach that predict the rate of an item by a user. Our method belongs to a pairwise learning to rank approach, such as [5].

ABCNN [20] is a convolution neural network with attention for question answering as a classification rather than reranking and shows high mean average precision. We can integrate attention mechanism into our methods for obtaining higher mean average precision.

6 Conclusion

We proposed a embeddings-based reranking with a boosting. Our embeddings-based reranking model learns to rank a relevant document higher with a similarity defined calculated word emmbeddings. Experimental results on English and Japanese Wikipedia showed that the proposed method outperformed SSI.

References

1. Bai, B., Weston, J., Grangier, D., Collobert, R., Sadamasa, K., Qi, Y., Chapelle, O., Weinberger, K.: Supervised semantic indexing. In: Proceedings of the 18th CIKM (CIKM 2009), pp. 187–196 (2009)
2. Berger, A., Lafferty, J.: Information retrieval as statistical translation. In: Proceedings of the 22nd SIGIR (SIGIR 1999), pp. 222–229 (1999)
3. Bespalov, D., Bai, B., Qi, Y., Shokoufandeh, A.: Sentiment classification based on supervised latent n-gram analysis. In: Proceedings of the 20th CIKM (CIKM 2011), pp. 375–382 (2011)
4. Blei, D., Ng, A., Jordan, M.: Latent dirichlet allocation. J. Mach. Learn. Res. **3**, 993–1022 (2003)
5. Burges, C., Shaked, T., Renshaw, E., Lazier, A., Deeds, M., Hamilton, N., Hullender, G.: Learning to rank using gradient descent. In: Proceedings of the 22nd International Conference on Machine Learning (ICML 2005), pp. 89–96 (2005)
6. Deerwester, S., Dumais, S.T., Furnas, G.W., Landauer, T.K., Harshman, R.: Indexing by latent semantic analysis. J. Am. Soc. Inf. Sci. **41**(6), 391–407 (1990)
7. Duchi, J., Hazan, E., Singer, Y.: Adaptive subgradient methods for online learning and stochastic optimization. Technical report UCB/EECS-2010-24. EECS Department, University of California, Berkeley, March 2010
8. Freund, Y., Iyer, R., Schapire, R.E., Singer, Y.: An efficient boosting algorithm for combining preferences. J. Mach. Learn. Res. **4**, 933–969 (2003)
9. Freund, Y., Schapire, R.E.: A decision-theoretic generalization of on-line learning and an application to boosting. J. Comput. Syst. Sci. **55**(1), 119–139 (1997)
10. Grangier, D., Bengio, S.: A discriminative kernel-based model to rank images from text queries. IEEE Trans. Pattern Anal. Mach. Intell. (TPAMI) **30**, 1371–1384 (2008)
11. Iwakura, T.: A boosted semi-markov perceptron. In: Proceedings of the 17th CoNLL, pp. 47–55 (2013)
12. Joachims, T.: Optimizing search engines using click through data. In: Proceedings of the 8th KDD (KDD 2002), pp. 133–142 (2002)
13. Kingma, D.P., Ba, J.: Adam: a method for stochastic optimization. CoRR abs/1412.6980 (2014)
14. Mikolov, T., Sutskever, I., Chen, K., Corrado, G.S., Dean, J.: Distributed representations of words and phrases and their compositionality. In: Burges, C.J.C., Bottou, L., Welling, M., Ghahramani, Z., Weinberger, K.Q. (eds.) Advances in Neural Information Processing Systems, vol. 26, pp. 3111–3119. Curran Associates, Inc. (2013)
15. Min, K., Zhang, Z., Wright, J., Ma, Y.: Decomposing background topics from keywords by principal component pursuit. In: Proceedings of the 19th ACM International Conference on Information and Knowledge Management (CIKM 2010), pp. 269–278 (2010)

16. Pennington, J., Socher, R., Manning, C.D.: GloVe: global vectors for word representation. In: Empirical Methods in Natural Language Processing (EMNLP), pp. 1532–1543 (2014)
17. Wang, Y., Sun, H., Zhang, R.: AdaMF: adaptive boosting matrix factorization for recommender system. In: Li, F., Li, G., Hwang, S., Yao, B., Zhang, Z. (eds.) WAIM 2014. LNCS, vol. 8485, pp. 43–54. Springer, Cham (2014). doi:10.1007/978-3-319-08010-9_7
18. Weston, J., Bengio, S., Usunier, N.: Large scale image annotation: learning to rank with joint word-image embeddings. In: European Conference on Machine Learning (2010)
19. Xu, J., Li, H.: AdaRank: a boosting algorithm for information retrieval. In: Proceedings of the 30th SIGIR (SIGIR 2007), pp. 391–398 (2007)
20. Yin, W., Schütze, H., Xiang, B., Zhou, B.: ABCNN: attention-based convolutional neural network for modeling sentence pairs. TACL **4**, 259–272 (2016)
21. Zeiler, M.D.: ADADELTA: an adaptive learning rate method. CoRR abs/1212.5701 (2012)

Improving Shard Selection for Selective Search

Mon Shih Chuang[✉] and Anagha Kulkarni[✉]

Department of Computer Science, San Francisco State University,
1600 Holloway Ave, San Francisco, CA 94132, USA
mchuang@mail.sfsu.edu, ak@sfsu.edu

Abstract. The *Selective Search* approach processes large document collections efficiently by partitioning the collection into topically homogeneous groups (*shards*), and searching only a few shards that are estimated to contain relevant documents for the query. The ability to identify the relevant shards for the query, directly impacts Selective Search performance. We thus investigate three new approaches for the shard ranking problem, and three techniques to estimate how many of the top shards should be searched for a query (shard rank cutoff estimation). We learn a highly effective shard ranking model using the popular learning-to-rank framework. Another approach leverages the topical organization of the collection along with pseudo relevance feedback (PRF) to improve the search performance further. Empirical evaluation using a large collection demonstrates statistically significant improvements over strong baselines. Experiments also show that shard cutoff estimation is essential to balance search precision and recall.

1 Introduction

Exhaustive Search, where the complete document collection is searched for every query, is commonly believed to provide superior performance than approaches that search only a subset of the collection. *Selective Search* [18] has challenged this belief by demonstrating comparable performance with Exhaustive Search. In rare cases, Selective Search has outperformed Exhaustive Search, but this has been limited to performance at early ranks. In this paper, we break through this limitation by demonstrating consistent and significant improvements at early ranks as well as at deeper ranks using Selective Search. To accomplish this we focus our efforts on improving accuracy of the algorithms that select the search space for each query. Our motivation comes from the observation that a search space that is as *pure* and as *complete* as possible (contains only and all the relevant documents for the query), has the best potential to improve both, search precision and recall.

At indexing time, the Selective Search approach [8,15,16] partitions the document collection into *shards* based on document similarity. The resulting shards are topically homogeneous, that is, documents about the same or related topics are in the same shard. As such, the relevant documents for a query are clustered together into one or few shards [27]. At query time, Selective Search leverages

© Springer International Publishing AG 2017
W.-K. Sung et al. (Eds.): AIRS 2017, LNCS 10648, pp. 29–41, 2017.
https://doi.org/10.1007/978-3-319-70145-5_3

this topical organization by restricting query processing to a select few shards. The problem of selecting the shards that are likely to contain all (or most) of the relevant documents for the query, is referred to as the *shard selection* problem. Shard selection is often formulated as a two-part problem where the first step is to *rank* the shards based on their estimated relevance, and second step is to select a *cutoff* that determines how many of the top ranked shards are searched for the query. We propose three approaches for the ranking problem that leverage topical organization of the collection, richer query representations, learning to rank framework, and pseudo relevance feedback. We also investigate three estimators for the cutoff problem, each of which makes use of the distribution of the shard scores assigned by the ranking algorithms. Using detailed empirical evaluation we demonstrate that the improved ranking and cutoff approaches enable Selective Search to significantly improve over Exhaustive Search, not just at early ranks but also at ranks as deep as 1000.

2 Related Work

Selective Search falls under the subfield of distributed information retrieval [5,23] where the goal is to search across multiple existing resources (aka shards), and to aggregate the results. One of the important research problems in distributed IR is that of ranking the resources based on the number of relevant documents they contain for the query. A long line of research has studied this problem. The proposed algorithms can be roughly categorized into two groups: Big-document approaches, which are based on term frequency statistics of the resources. Small-document approaches, which are based on a small sample of documents from each resource.

Big-document approaches: In cooperative environments, the providers of target resources are willing to share the term statistics and metadata of the documents collections. The methods ranking resources based on the term statistics are called big-document approaches since the model can be extended from document retrieval model by treat each resource as a entity. GlOSS [12], and its extended versions gGloss [11] and vGlOSS [13] provide a solution of text database discovery problem. The algorithms uses the ratio of term frequency and database size, also other metadata like field information (title, body, links) for choosing the candidate resources. CORI [4,5] algorithm compiles document frequency (df), and shard frequency (sf) of each term, and computes the shard scores by extending the tf.idf formula as follows: $Score(t_k|S_i) = b + (1 - b) \times T \times I$, where t_k is the kth term in the query, S_i is the shard being scored, b: is a belief parameter that is set to the default value of 0.4, and T and I are defined as: $T = df_i/(df_i + 50 + 150 \times sw_i/avg\_sw)$, $I = \log(\frac{S+0.5}{sf})/\log(S+1.0)$, where df_i is the document frequency of the term t_k in shard S_i, sf is the shard frequency of term t_k, sw_i is the number of total words in the shard i, avg_sw is the average number of words in a shard, and S is the number of shards. Taily [1] is another big document approach. According to Kanoulas et al.'s work [14], the term frequency based document score across whole collection can be modeled by gamma curve distribution. Thus, Taily pre-computes two parameters scaler θ, and K of gamma distribution to fit the document score for every single-term query against every shard. By storing the score distribution of single-term query against every shard, it can estimate the score distribution of user query with multiple terms.

Small-document approaches: In uncooperative environments, the important statistics such as term frequency can not be obtained. Therefore, big-document approaches are not capable to compute the shard scores. Small-document algorithms solve this issue by approximating the document distribution inside a resource by sampling a small subset, which is called centralized sample database, or centralized sample index (CSI) structure. ReDDE [21,24] algorithm runs the user query against the CSI, and assumes that the top n retrieved documents are relevant. The original version of ReDDE compute a score for each shard as follows: $Score(S_i^q) = Count(n, S_i^q) \times (|S_i|/|S_i^{CSI}|)$, where $Count(n, S_i^q)$ is the count of documents occurred in top n retrieved documents in CSI $|S_i|$ is the size of the shard and $|S_i^{CSI}|$ is the size of its sample. The shard scores are then normalized to obtain a valid probability distribution used to rank the shards. CRCS [22] passes the user queries to CSI and compute the score of each resource from the returned document rank. Two version of CRCS are introduces by modifying the function of rank. CRCS(1) uses a simple linear decreasing model, and CRCS(e) uses an exponential decaying model for the document score. SUSHI [26] passes the user queries to CSI and uses the returned document scores to estimate the score distribution. For each shard, SUSHI fits one of three types of distribution curves, linear, logarithmic, and exponential to the scores of returned documents in CSI.

3 Shard Ranking

We investigate three different shard ranking approaches that explore improved query representation, and ranking algorithms.

3.1 CORI_Uni+Bi and ReDDE_Uni+Bi

The setup in which Selective Search operates is cooperative, that is, CORI has access to the complete term statistics for each shard. While ReDDE estimates the shard ranking based on the small sample of each shard. As such, one would expect CORI to perform at least on par, if not outperform, ReDDE. However, the term statistics that CORI uses, although complete are highly misleading. This is so because traditional CORI uses unigram model where each query term is treated as an individual entity. This leads to a very poor representation of the query. For example, the query *arizona fish and game* when decomposed into unigrams, looses the central topic of *fishing in arizona*.

These observations motivate the following investigation where a richer representation of the query is used by both CORI and ReDDE. Given a query with n terms, $n-1$ bigrams are generated by enumerating all pairs of consecutive terms.

Bigrams with stopwords are discarded, and the remaining bigrams are added to the original unigram query. As an example, for query *obama family tree*, this approach generates the following query representation using the Indri Query Language: *#combine(obama family tree #uw2(obama family) #uw2(family tree))*, where the *#combine* operator coalesces the scores from all the element of the query, and the *#uwX* operator is used for specifying an unordered phrase of length X. The ReDDE_Uni+Bi runs the query generated using the above procedure against the CSI, and the rest of the search process is same as before. In case of CORI_Uni+Bi, frequency statistics for bigrams, in addition to unigrams, are used in order to evaluate the richer query representation. Since the bigram statistics can be precomputed off-line, the query response time is not affected.

3.2   LeToR-S (Learning to Rank Shards)

The Learning-to-Rank (LTR) approaches have been successfully used to improve the document ranking task [3,10,28]. Recently, LTR approaches have been applied to the problem of shard ranking [6,8], but further exploration is needed. In this work we investigate if a ranking model can be learned for the shard ranking task. To test this intuition we start by defining the set of features for the ranking model, and the target variable. The number of relevant documents in the shard for the query, is used as the target. Model features are described next.

LeToR-S Features: The *title, body, heading, url, whole document*, each generates a separate set of features. Three variants of CORI scores, *CORI_SUM*, *CORI_MIN*, and *CORI_VAR* (given below), are computed by evaluating the query against the above five fields. Furthermore, two query representations are evaluated:- unigram, and phrasal. As such, this generates 30 distinct features.

$$CORI\_SUM(Q|S_i) = \sum_{t_k \in Q} Score(t_k|S_i) \tag{1}$$

$$CORI\_MIN(Q|S_i) = \min_{t_k \in Q} Score(t_k|S_i) \tag{2}$$

$$\mu = \frac{1}{n} \sum_{t_k \in Q} Score(t_k|S_i) \tag{3}$$

$$CORI\_VAR(Q|S_i) = \sum_{t_k \in Q} (Score(t_k|S_i) - \mu)^2 \tag{4}$$

where $Score(t_k|S_i)$ is the score of one n-gram term t_k computed using the standard CORI formulation (Sect. 2). For every query and shard pair, the above vector of 30 features is computed, and a ranking model is learned using the RandomForest [2] algorithm implemented in the RankLib [9] library from the Lemur Project [7].

3.3 LeToR-SWP (LeToR-S with Wikipedia-Based Pseudo Relevance Feedback)

One of the primary challenges for the shard ranking algorithm is posed by the paucity of information in the query. The query length is typically 2 to 3 words, and many queries even contain just one term. Many shard ranking errors can be attributed to the shortness of queries. To address this problem we turn to a well-studied solution used for document ranking, *pseudo relevance feedback* (PRF). However, to avoid the typical problems with PRF, such as, *topic drift*, we develop an approach that uses PRF in a controlled fashion as follows. As a pre-processing step, a *profile* for every shard is compiled, which consists of the top 1000 terms with highest document frequency (DF) in the shard. As an example, the top 5 terms in the profile of one of the shards are *rum piano classical acoustic composer*, which are all related to music, while for another shard the top profile terms are *pepper butter onion chop fry*, which are all related to cooking. At query time, the LeToR-S approach is employed to obtain a shard ranking for the original query. The *profile* of the topmost ranked shard is used to identify the expansion candidates. Specifically, 100 terms with highest $DF.ISF$ are selected, where $DF.ISF$ is defined as: $DF.ISF(t, S_i) = \log(1 + DF(t, S_i)) \times \log(ISF(t))$, where $DF(t, S_i)$ is the document frequency of the term t in the shard S_i, and ISF is: $ISF(t) = M/\sum_{i=1 to M} I(t, S_i)$ where M is the total number of shards, and I is an indicator function that returns 1 if the term t occurs in the *profile* of shard S_i, and 0 otherwise.

The expansion candidate terms are further filtered using Wikipedia as follows. The query is run against the English Wikipedia (https://en.wikipedia.org/w/api.php) to obtain the summary section of the Wikipedia entry. If this search leads to a Wiki disambiguation page then the expansion process is aborted for that query. Only those candidate terms that occur in the retrieved Wikipedia summary are added to the query to generate the expansed version. The expanded query is then employed to obtain a revised shard ranking using the LeToR-S approach. This query expansion based shard ranking process is only conducted for single-term queries since that is where the paucity of information is most stark. We refer to this overall approach as LeToR-SWP.

4 Dynamic Shard Rank Cutoff Estimation

When more than necessary shards are searched for a query, it degrades the search efficiency, and the search effectiveness often degrades too because the likelihood of false positive documents getting retrieved increases. At the same time, searching fewer than necessary shards definitely hurts search effectiveness. As such, estimating the optimal number of top ranked shards that should be searched for the query is an important problem [1,17,20]. We refer to it as *shard rank cutoff estimation*. This estimation needs to be query-specific since the optimal cutoff differs across queries based on the topical spread of the queries. We investigate the impact of three different shard rank cutoff estimators on the overall search effectiveness, by contrasting it with the standard approach of selecting a fixed rank cutoff for all queries.

The first two cutoff estimation methods attempt to locate the *elbow* pattern in the shard scores showed by Fig. 2. Typically, the shard scores computed for a query by a shard ranking algorithm, decrease rapidly at early ranks and then plateau off. The rank at which the transition to plateau occurs is often the optimal cutoff. The PK2 approach [19] compares two consecutive shard scores to locate the elbow. Specifically, PK2 computes the ratio of shard scores of the current rank and the previous rank. The mean and one standard deviation of the PK2 scores is then computed, and the rank at which the PK2 value is greater than the upper bound of the standard deviation range is selected as the cutoff rank. This ensures that points on the plateau, and the points on the falling edge of the elbow are not selected, and only the ones on the elbow are selected. The second approach for cutoff estimation, PK3, compares three consecutive scores to locate the elbow. PK3 is defined as: $PK3(r) = \frac{2 \times Score(r)}{Score(r-1) + Score(r+1)}$. Like for PK2, the mean and one standard deviation of PK3 values is computed and the rank at which the PK3 value is greater than the upper bound of the standard deviation is selected as the cutoff rank. For both, PK2 and PK3, the number of data points (M) over which the mean and the standard deviation are computed is a tunable parameter.

The third cutoff estimation approach is a variant of the Rank-S algorithm [20], where the distribution of the shard scores is used to estimate the cutoff rank. Shard scores that drop rapidly along the ranks, suggest early cutoff, whereas shard scores that reduce slowly, indicate deeper cutoff. Based on this intuition an exponential decay function is applied to the shard scores, as follows: Rank-$S(r) = Score(r) \times B^{-r}$, where r is the rank, and B is a tunable parameter to control the decaying rate. The rank at which Rank-S converges to 0 (Rank-S $\leq$ 0.0001) is the cutoff estimate.

5   Empirical Evaluation

For the empirical evaluation we use the CategoryB dataset of ClueWeb09, which contains 50,220,423 documents. The 92 topical shards created for CategoryB dataset by Kulkarni and Callan [20] are used in this work. The evaluation queries are from TREC Web Track 2009–2012. Out of the 200 queries, 6 queries do not contain any relevant document in this dataset, and thus are discarded. The remaining 194 queries are divided into 10-fold for the LeToR-S experiment to facilitate 10-fold cross validation. For the small-document approach, ReDDE, we construct the CSI by randomly sampling 0.5% of the documents from every shard. The search engine used in our experiment is Indri 5.9 [25] from Lemur Project. For all the Selective Search experiments reported in this sections, the top 10 shards were searched for each query. This corresponds to a search space of about 5.5 million documents. This is an order of magnitude smaller than the search space of Exhaustive Search (50+ million documents). Table 1 presents a results for four different shard ranking investigations that we undertook. Statistical significance testing was performed using paired T-test at $p < 0.05$.

Table 1. Exhaustive Search, and Selective Search with different shard ranking algorithms. $\dagger^i$: statistically significant improvements when compared to approach in row i. $\triangledown$: significantly worse values when compared to Exhaustive Search.

#	Run	P@30	P@100	MAP@1000	R@30	R@100	NDCG@1000
1	Exhaustive	0.254	0.189	0.181	0.174	0.374	0.429
2	CORI	0.255	0.178$\triangledown$	0.164$\triangledown$	0.158	0.329$\triangledown$	0.370$\triangledown$
3	ReDDE	0.256	0.182$\triangledown$	0.172$\triangledown$	0.175	0.360$\dagger^2$	0.400$\dagger^2\triangledown$
4	CORI_Uni+Bi	0.271$\dagger^{1,2}$	0.188$\dagger^2$	0.181$\dagger^2$	0.180$\dagger^2$	0.363$\dagger^2$	0.408$\dagger^2\triangledown$
5	ReDDE_Uni+Bi	0.263	0.184	0.175	0.182$\dagger^2$	0.363$\dagger^2$	0.398$\dagger^2\triangledown$
6	LeToR-S	0.270$\dagger^{1,2}$	0.194$\dagger^{2..5}$	0.186$\dagger^{2..5}$	0.179$\dagger^2$	0.377$\dagger^{2..5}$	0.417$\dagger^{2..5}\triangledown$
7	LeToR-SWP	0.281$\dagger^{1..6}$	0.197$\dagger^{1..5}$	0.191$\dagger^{1..5}$	0.186$\dagger^{1..6}$	0.382$\dagger^{2..5}$	0.423$\dagger^{2..5}$

5.1 Big-Document Versus Small-Document Shard Ranking Approach

The goal of the first experiment was to confirm the conventional belief that *small-document approaches provide superior search effectiveness than big-document approaches, specifically in the context of topical shards*. The first three rows in Table 1 are the focus of this analysis. As compared to Exhaustive Search, CORI and ReDDE, both struggle at deeper ranks. CORI, the big document approach, is consistently inferior to ReDDE, the small document approach, across all the metrics. In fact, at deeper ranks (R@100 and NDCG@1000) the improvements over CORI, with ReDDE are statistically significant. These results confirm that the conventional unigram language model based shard ranking approach adopted by CORI struggles to differentiate the relevant shards from non-relevant shards. This is so even for topical shards where the distribution of relevant documents across shards is highly skewed. Also, note that CORI has access to the vocabulary of the complete collection whereas ReDDE is only using 0.5% subset of the collection for estimating the shard ranking. Thus CORI's inferior performance is especially surprising. These observations motivate the experiment described next.

5.2 Effect of Richer Query Representation

CORI's inferior performance with multi-term queries motivates the investigation in this section. The results with CORI and ReDDE when using the richer query representation are given in rows 4 and 5 of Table 1. These results show an opposite trend as that with unigram query representation (rows 2 and 3). The big-document approach (CORI_Uni+Bi) performances better, although not significantly, than the small-document approach (ReDDE_Uni+Bi). CORI clearly benefits more from the richer query representation than ReDDE. CORI_Uni+Bi results are significantly better than those with CORI. This is not the case for ReDDE_Uni+Bi. At early ranks, CORI_Uni+Bi is significantly better than even Exhaustive Search. This indicates substantial reduction in false-positives at early ranks. This is facilitated by two factors:- topic-based shards reduce the noise in each shard, and CORI_Uni+Bi selects the shards such that the resulting search space is *purer* than that used by Exhaustive Search.

5.3   Learning to Rank Shards (LeToR-S)

The 6th row in Table 1 reports the results with Learning to Rank Shards approach (LeToR-S). The first obvious trend in these results is that LeToR-S significantly outperforms the current best, CORI_Uni+Bi, at deeper ranks. At early ranks, however, the two approaches provide comparable precision and recall. To understand these results better we analyze a few queries in detail. For one of the queries, *getting organized*, CORI_Uni+Bi ranks the shard with most number of relevant documents at 11th position, while LeToR-S ranks it at 3rd. This is a difficult query because both the query terms are common terms. Even when the terms are treated as a phrase, it is still not a focused query. This is reflected in the low scores assigned to relevant shards by CORI_Uni+Bi. LeToR-S, however, uses meta-data in addition to the document contents for defining the features. One particular meta-data feature, *url* field, proves to be especially valuable for this query. In short, LeToR-S benefits from having the field score features, while CORI_Uni+Bi suffers because it only uses document contents for shard ranking. For queries, *battles in the civil war* and *kansas city mo*, as well, CORI_Uni+Bi ranks the most relevant shard at a much deeper rank than LeToR-S.

Another feature category that helps LeToR-S outperform CORI_Uni+Bi is the CORI minimum score features. Recall that the CORI minimum score feature is lowest CORI score computed for the individual query terms against a shard. This feature models the intuition that all of the query terms should have high CORI score for a relevant shard. Low CORI score, even if only for one of the query terms, indicates less likelihood of shard relevance. For query, *pacific northwest laboratory* only one shard contains all the relevant documents, LeToR-S ranks this shard at 8th place, while CORI_Uni+Bi ranks it at 11. Through the CORI minimum score feature, 3 false-positive shards are eliminated by LeToR-S from the shard ranking. These false-positive shards have high overall CORI score because some of the query terms have high CORI score, and thus dominate the cumulative score. However, the CORI minimum score captures that some query terms have low CORI score for these false-positive shards and thus push them down in the shard ranking. The results in Table 1 also indicate that at early ranks LeToR-S performs significantly better than Exhaustive Search. This improvement often comes from single term ambiguous queries (*euclid, avp, iron, unc*). The topic-based partitioning of the collection organizes the documents with similar meaning into the same shard. Often one of the meanings is more dominant than others in the collection, that is also often the relevant meaning for the query. Shards with the dominant meaning have higher document frequency (*df*) than shards with the rare meaning, and thus documents with dominant meaning only are searched. This reduces the false-positive documents (documents with rare meaning) from the result, and thus improves the search precision.

5.4 LeToR-SWP

The last row in Table 1 reports the results for shard ranking approach with PRF described in Sect. 3.3. Recall that we apply this approach only to single term queries in order to minimize the risks involved with applying PRF, and because single term queries need expansion the most. Out of the 194 queries, 52 are single term queries, out of which 34 have a Wikipedia entry. Furthermore only 26 out of these 34 queries get expanded because for the remaining queries, the *profile* of the top ranked shard does not contain any term that also occurs in the corresponding Wikipedia summary. Although, this is a highly restrictive approach and only expands a small fraction of the queries, it provided consistent improvements in performance, as evidenced by the results in Table 1. In fact, at early ranks the improvements over a strong baseline, LeToR-S, are statistically significant.

Using this technique one of the queries, *iron* is expanded to *iron powder weld carbon temperature* and leads to 118% improvement over Exhaustive Search, in MAP@1000. Another query, *joints* is expanded to *joints brain bone*, *sat* is expended to *sat academy admission mathematics assessment*. Many of these single term queries are ambiguous, and the expansion resolves the ambiguity by picking the dominant sense. Some queries receive more than 10 terms through the expansion process, for instance, query *starbucks* expanded to *starbucks taste snack chip juice bean cream beverage fresh tea serving coffee* and improves 10% over Exhaustive Search in MAP@1000 metric. Overall, the performance of LeToR-SWP is 11%, 4%, 6%, 7%, 2% higher than that with Exhaustive Search in P@30, P@100, MAP@1000, R@30, and R@100, respectively. The only metric that is not improved with LeToR-SWP is NDCG@1000.

Table 2. Exhaustive Search and Selective Search with LeToR-SWP, and different shard rank cutoff approaches. Precision-oriented optimization and evaluation. AvgT: Average shard rank cutoff. †: statistically significant improvements as compared to Fixed Cutoff of 10. ‡: statistically significant improvements over Exhaustive Search.

Cutoff Approach	AvgT	P@30	MAP@30	R@30	NDCG@30
Exh	92	0.254	0.071	0.174	0.215
Fixed	10	0.281‡	0.084‡	0.186‡	0.244‡
PK2 M = 20	7	0.289†‡	0.089†‡	0.187	0.252†‡
PK3 M = 20	8	0.285‡	0.084‡	0.181	0.244‡
Rank-S B = 3	8	0.283‡	0.085‡	0.187	0.245‡

5.5 Dynamic Shard Rank Cutoff Estimation

All of the Selective Search experiments reported until now used a fixed shard rank cutoff of 10. Tables 2 and 3 report the results with dynamic shard rank cutoff estimators. The two baselines, Exhaustive Search, and Selective Search with fixed rank cutoff of 10, are given in the first two rows of the tables. For all of the Selective Search experiments the same shard ranker was used, LeToR-SWP. Table 2 provides the *precision-oriented* experiments where the cutoff estimators were tuned to optimize search performance at early ranks. The results for the corresponding recall-oriented experiments are in Table 3. This separate optimization of cutoff estimators is necessary because much fewer shards have to be searched for a precision-oriented task than for a recall-oriented task. This is evidenced by the values reported in the second column (AvgT) which provides the average shard cutoff predicted by the different approaches. Also, an upper bound of 10 was enforced on the estimators for the precision-oriented experiments, and 50 for the recall-oriented experiments.

Table 3. Exhaustive Search and Selective Search with LeToR-SWP, and different shard rank cutoff approaches. Recall-oriented optimization and evaluation. †: statistically significant improvements as compared to Fixed Cutoff of 25. ‡ and ▽: statistically significant improvement or degradation over Exhaustive Search, respectively.

Cutoff approach	AvgT	P@1000	MAP@1000	R@1000	NDCG@1000
Exh	92	0.042	0.181	0.755	0.429
Fixed	25	0.041▽	0.185‡	0.728▽	0.427
PK2 M = 92	26	0.041	0.191†‡	0.731▽	0.432
PK3 M = 92	25	0.041	0.189‡	0.727▽	0.429
Rank-S B = 1.3	22	0.041	0.187‡	0.724▽	0.428

The results in Table 2 show that all the Selective Search experiments with cutoff estimators perform significantly better than Exhaustive Search. Even though the PK2 experiment is searching fewer shards (7) than the fixed cutoff setting (10), the corresponding P@30, MAP@30, and NDCG@30 values are significantly higher than those with fixed cutoff. By searching fewer shards the false-positive documents in the final results are reduced. The improvements with PK2 over Exhaustive Search are 14%, 25%, 7%, 17% in P@30, MAP@30, R@30, and NDCG@30, respectively. The recall-oriented results in Table 3 show that performance on P@1000, and NDCG@1000 with PK2 is comparable to both the baselines, but R@1000 still remains a challenge. However, PK2 improves one of the toughest metrics, MAP@1000, where the performance is significantly better than both, Exhaustive Search, and Fixed cutoff Selective Search. This result is a first of its kind.

The distribution of rank cutoffs predicted by PK2, PK3, and Rank-S for the two setting of precision-oriented and recall-oriented are given in Fig. 1. The prominent trend in these distributions is the differences in the skew of the predictions. PK2 predictions are substantially less skewed than those with the other two predictors. PK3 estimates 10 or higher cutoff for majority of the queries in both the experimental settings. Although, the upper bound restriction limits

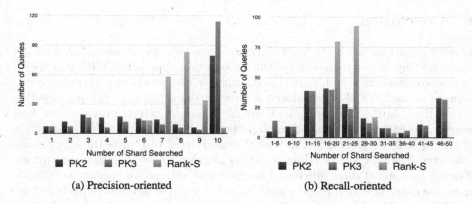

Fig. 1. Distribution of shard rank cutoff predictions by PK2, PK3, and Rank-S.

PK3 predictions to top 10 shards in the precision-oriented experiment. PK2 also estimates a cutoff 10 or higher for about half of the queries. This is because the shard ranking is not perfect, several false positive shards still get ranked above the true positive shards. As such, the elbow point occurs at a deeper rank than that in an ideal shard ranking (Fig. 2). The predictions with Rank-S demonstrate narrow spread in both the settings which is a function of the rapid decay in shard scores cause by its formulation. In recall-oriented setting the range of the predictions for PK2 and PK3 is much large, which indicates the location of the elbow varies substantially from one query to another. This underscores the necessity of dynamic shard rank cutoff predictor. Overall, PK2 appears to have the lowest bias, and the effects of that are apparent in the search results.

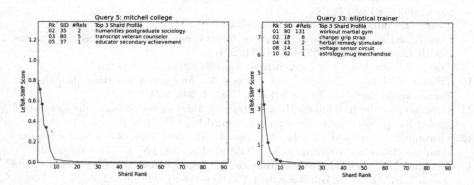

Fig. 2. *Elbow* formed by LeToR-SWP shard scores. Blue dots represent the shards that contain relevant documents, and the text shows the ranks and profiles of those shards. (Color figure online)

6   Conclusion

Our goal for this work was to investigate ways to improve shard selection for Selective Search. We revived an old shard selection approach, CORI, which supported competitive search performance when it was provided with richer query representation. We also introduced a novel shard ranking algorithm based on the well-established Learning-To-Ranking approach, which provided the best search precision while also sustaining the recall. To address data paucity in short queries, we leveraged the topical organization of shards through pseudo relevance feedback and Wikipedia, which improved both, precision and recall. Finally, we applied three different dynamic cutoff estimation approaches to optimize the shard selection process further, which resulted in improvements as high as 25% at early ranks, and small but consistent improvements at deeper ranks.

References

1. Aly, R., Hiemstra, D., Demeester, T.: Taily: shard selection using the tail of score distributions. In: Proceedings of the SIGIR Conference, pp. 673–682. ACM (2013)
2. Breiman, L.: Random forests. Mach. Learn. **45**(1), 5–32 (2001)
3. Burges, C.J.C.: From ranknet to lambdarank to lambdamart: an overview. Learning **11**(23–581), 81 (2010)
4. Callan, J., Lu, Z., Croft, B.: Searching distributed collections with inference networks. In: Proceedings of the SIGIR Conference, pp. 21–28. ACM (1995)
5. Callan, J.: Distributed information retrieval. In: Croft, W.B. (ed.) Advances in Information Retrieval. The Information Retrieval Series, vol. 7, pp. 127–150. Springer, Boston (2002). doi:10.1007/0-306-47019-5_5
6. Chuang, M.-S., Kulkarni, A.: Balancing precision and recall with selective search. In: Proceedings of 4th Annual International Symposium on Information Management and Big Data (2017)
7. Croft, B., Callan, J.: Lemur project (2000)
8. Dai, Z., Kim, Y., Callan, J.: Learning to rank resources. In: Proceedings of the SIGIR Conference, pp. 837–840. ACM (2017)
9. Dang, V.: Lemur project components: Ranklib (2013)
10. Freund, Y., Iyer, R., Schapire, R.E., Singer, Y.: An efficient boosting algorithm for combining preferences. J. Mach. Learn. Res. **4**, 933–969 (2003)
11. Gravano, L., Garcia-Molina, H.: Generalizing gloss to vector-space databases and broker hierarchies. Technical report, Stanford InfoLab (1999)
12. Gravano, L., Garcia-Molina, H., Tomasic, A.: The effectiveness of gioss for the text database discovery problem. ACM SIGMOD Rec. **23**, 126–137 (1994). ACM
13. Gravano, L., García-Molina, H., Tomasic, A.: Gloss: text-source discovery over the internet. ACM Trans. Database Syst. (TODS) **24**(2), 229–264 (1999)
14. Kanoulas, E., Dai, K., Pavlu, V., Aslam, J.A.: Score distribution models: assumptions, intuition, and robustness to score manipulation. In: Proceedings of the SIGIR Conference, pp. 242–249. ACM (2010)
15. Kim, Y., Callan, J., Culpepper, J.S., Moffat, A.: Does selective search benefit from WAND optimization? In: Ferro, N., et al. (eds.) ECIR 2016. LNCS, vol. 9626, pp. 145–158. Springer, Cham (2016). doi:10.1007/978-3-319-30671-1_11

16. Kim, Y., Callan, J., Culpepper, J.S., Moffat, A.: Efficient distributed selective search. Inf. Retr. J. **20**(3), 221–252 (2017)
17. Kulkarni, A.: ShRkC: shard rank cutoff prediction for selective search. In: Iliopoulos, C., Puglisi, S., Yilmaz, E. (eds.) SPIRE 2015. LNCS, vol. 9309, pp. 337–349. Springer, Cham (2015). doi:10.1007/978-3-319-23826-5_32
18. Kulkarni, A., Callan, J.: Selective search: efficient and effective search of large textual collections. ACM Trans. Inf. Syst. (TOIS) **33**(4), 17 (2015)
19. Kulkarni, A., Pedersen, T.: How many different "john smiths", and who are they? In: AAAI, pp. 1885–1886 (2006)
20. Kulkarni, A., Tigelaar, A.S., Hiemstra, D., Callan, J.: Shard ranking and cutoff estimation for topically partitioned collections. In: Proceedings of the CIKM Conference, pp. 555–564. ACM (2012)
21. Markov, I., Crestani, F.: Theoretical, qualitative, and quantitative analyses of small-document approaches to resource selection. ACM Trans. Inf. Syst. (TOIS) **32**(2), 9 (2014)
22. Shokouhi, M.: Central-rank-based collection selection in uncooperative distributed information retrieval. In: Amati, G., Carpineto, C., Romano, G. (eds.) ECIR 2007. LNCS, vol. 4425, pp. 160–172. Springer, Heidelberg (2007). doi:10.1007/978-3-540-71496-5_17
23. Shokouhi, M., Si, L., et al.: Federated search. Found. Trends® Inf. Retr. **5**(1), 1–102 (2011)
24. Si, L., Callan, J.: Relevant document distribution estimation method for resource selection. In: Proceedings of the SIGIR Conference, pp. 298–305. ACM (2003)
25. Strohman, T., Metzler, D., Turtle, H., Croft, W.B.: Indri: a language model-based search engine for complex queries. In: Proceedings of the International Conference on Intelligent Analysis, vol. 2, pp. 2–6. Citeseer, 2005
26. Thomas, P., Shokouhi, M.: Sushi: scoring scaled samples for server selection. In: Proceedings of the SIGIR Conference, pp. 419–426. ACM (2009)
27. van Rijsbergen, C.J.: Information Retrieval. Butterworth, London (1979)
28. Xu, J., Li, H.: Adarank: a boosting algorithm for information retrieval. In: Proceedings of the SIGIR Conference, pp. 391–398. ACM (2007)

IR Models and Theories

Learning2extract for Medical Domain Retrieval

Yue Wang[✉], Kuang Lu, and Hui Fang

Department of Electrical and Computer Engineering,
University of Delaware, Newark, USA
{wangyue,lukuang,hfang}@udel.edu

Abstract. Search is important in medical domain. For example, physicians need to search for literature to support their decisions when they diagnose the patients, especially for the complicated cases. Even though they could manually input the queries, it is not an easy task because queries are expected to include enough information about the patients. Therefore, the queries tend to be verbose. However, those verbose queries may not work well since the search engine would favor documents covering every term in the query, but not the ones which are truly important. Existing work on verbose query processing in Web search has studied the similar problem, but the methods are not applicable to the medical domain because of the complexity of the medical queries and the lack of domain-specific features. In this work, we propose a set of new features to capture the importance of the terms which are helpful for medical retrieval, i.e., **Key Terms**, from verbose queries. Experiment results on the TREC Clinical Decision Support collections show that the improvement of using the selected Key Terms over the baseline methods is statistically significant.

1 Introduction

Medical records contain valuable resources, such as the diagnoses and treatments, for the patients. In recent years, the growing usage of Electronic Medical Records (EMR) makes it possible for the physicians to access this valuable resource. One notable search scenario is, before the physicians make the clinical decisions, they need to browse previous medical records and literature that are similar to the situation of the current patient in order to ensure the accuracy of the diagnose, test, or treatment they would provide to the patient, especially for the complicated cases. Although the physician can manually enter the queries, these queries often need to be deliberated to ensure the search quality, since there are lots of detailed information about the patient, and it is not straightforward to identify which information should or should not to be included in the search query. Table 1 shows an example of how queries are formulated based on the EMR. Since the current search engines assume the queries are composed by key words, the documents that cover more query terms would be favored. However, from the example in Table 1, it is clear that not every term in the EMR is equally important. Returning the documents containing fewer important terms would not be helpful for the physicians. Thus, useful terms selection from the

W.-K. Sung et al. (Eds.): AIRS 2017, LNCS 10648, pp. 45–57, 2017.
https://doi.org/10.1007/978-3-319-70145-5_4

Table 1. An example showing how queries are formulated based on the EMR

Query type	Query content
The EMR of the patient	78 M w/ pmh of CABG in early [**Month (only) 3**] at [**Hospital6 4406**] (transferred to nursing home for rehab on [**12-8**] after several falls out of bed.) He was then readmitted to [**Hospital6 1749**] on [**3120-12-11**] after developing acute pulmonary edema/CHF/unresponsiveness? There was a question whether he had a small MI; he reportedly had a small NQWMI. He improved with diuresis and was not intubated. Yesterday, he was noted to have a melanotic stool earlier this evening and then approximately 9 loose BM w/ some melena and some frank blood just prior to transfer, unclear quantity
A shorter version of the EMR	78 M transferred to nursing home for rehab after CABG. Reportedly readmitted with a small NQWMI. Yesterday, he was noted to have a melanotic stool and then today he had approximately 9 loose BM w/ some melena and some frank blood just prior to transfer, unclear quantity
Simplified query	A 78 year old male presents with frequent stools and melena

EMR becomes an essential but challenging task, even for the physicians with extensive medical knowledge.

This problem is similar to verbose query processing in the Web search. Although the information retrieval with verbose query has been studied, existing methods are not applicable in the medical domain for two reasons. On one hand, existing work considered the queries with 5 or more terms as verbose queries [1], but the queries in medical domain are much longer and much more complicated than the Web queries. This can be clearly observed from the example in Table 1 as the simplified query still contains 11 terms. On the other hand, the features selected for the web queries may not work for the medical domain. For instance, the features which require the query logs of the general search engine, such as query log frequency [2] and similarity with previous queries [3], etc., can not be directly used in the medical domain because of lack of history of the verbose queries.

In this paper, to overcome the comprehensive requirement of the medical related knowledge for simplifying the verbose queries, we propose an automatic way to extract useful information from the verbose queries in the medical domain. Specially, we designed a set of features which could be helpful for identifying the Key Terms in the verbose queries. We then applied state-of-art machine learning techniques with the proposed features to select the terms used for retrieval. The experimental results over the TREC CDS collections showed that the proposed features could improve the performance.

2   Related Work

We define our work as key terms identification in medical domain. There are several research areas that are related to our work.

Clinical Query Generation. Soldaini et al. studied the query reduction techniques for searching medical literature [4]. They followed the structure proposed in [5], which takes quality predictors as features to rank the sub-queries of the original query using SVM. In addition, they also studied how to utilize query expansion technique with the query reduction. However, they did not report the performance of their method on the verbose queries of TREC CDS collection: although they also used the public available CDS data collection, they created their own query set to test the performance.

Keyphrase Extraction. The concept of identifying useful information from verbose query was introduced by Turney [6]. After that, considerable amount of works have been done in this area [1,3,5]. Bendersky and Croft proposed to identify key concepts from given verbose queries using a set of features [2]. They considered the key concepts in a verbose query as a special form of sub-queries, and then proposed to use the machine learning methods to predict the usefulness of all the sub-queries. The experiments are conducted using the standard TREC collections (Robust04, W10G, and Gov2). Our work is similar to theirs, however, the differences are also clear. On one hand, our verbose query are much longer than the ones used in their experiments. On the other hand, we focus on the medical domain, so we would also like to explore how the domain specific features could be used in our experiments.

Medical Domain Retrieval. Bio-medical domain retrieval has received more and more attentions in the recent years. Existing work could be divided into two directions based on how the documents are represented, i.e., term based representation and concept based representation. The term based representation adopted the traditional bag-of-term assumption which consider each term independently. They then applied other techniques, such as query expansion with domain resources [7,8], semantic similarity of the documents and the corresponding pseudo relevance feedback set [9], or a combination of different retrieval models [10] and types of documents [11] to improve the retrieval performance.

Concept-based representation assumes the documents are composed by concepts. It relies on specific NLP toolkits, such as MetaMap or cTAKEs, to identify the concepts from the raw documents and then apply the existing retrieval methods [12–15]. Wang and Fang showed that the results from the NLP toolkit could generate less satisfied results because the concepts from the same aspect are related, and the one-to-many mapping from the MetaMap could inflate the weights of some query aspects [16]. In order to solve this problem, they proposed two weighting regulations to the existing retrieval models. Despite the different representation methods, the queries used in these work are the simplified version of the EMR, which is different from our problem since we focus on how to select the important terms from the verbose query in medical domain.

3   Methods

We define the terms that could be helpful for retrieving relevant documents in medical domain as **Key Terms**. The goal of our research is to identify those Key Terms from the verbose queries. We formulate this problem as a classification problem. Formally, the input of our system is the query $\mathbb{Q}$ which contains n terms, i.e., $\mathbb{Q} = (t_1, t_2, \cdots, t_n)$. The classification problem is then to infer a Key Term label for each term, i.e., for each term t_i, to classify whether it is a Key Term. The Key Terms would be kept and then used for retrieval propose.

Since we model this problem as a classification problem, the feature selection is the key component to success. In this work, we propose several new features for this domain specific problem, as well as adopt some features from existing work [2]. The list of all the features is included in the Table 2. Due to the limited space, we will only introduce the important ones in the following discussion.

Table 2. Features used for identifying Key Terms

Type	Feature	Description
Domain features	$Concept(t_i)$	Whether t_i is part of a medical concept
	$Unique(t_i)$	The ratio of the IDF value of t_i in medical and web domain
Lexicon features	$Abbr(t_i)$	Whether t_i is an abbreviation
	$All\_Cap(t_i)$	Whether t_i only contains capital letters
	$Capitalized(t_i)$	Whether t_i contains any capital letters
	$Stop(t_i)$	Whether t_i is a stopword
	$Numeric(t_i)$	Whether t_i is a number
POS features	$Noun(t_i)$	Whether t_i is a noun, or part of a noun phrase
	$Verb(t_i)$	Whether t_i is a verb, or part of a verb phrase
	$Adj(t_i)$	Whether t_i is an adjective
Statistical features	$tf_{des}(t_i)$	The term frequency in description of t_i
	$tf_c(t_i)$	The term frequency in collection of t_i
	$IDF(t_i)$	The invert document frequency t_i
	$wig(t_i)$	The weighted information gain of t_i (Proposed in [2])
Locality features	$Rank_{des}(t_i)$	The position of t_i shown in the description
	$Rank_{sent}(t_i)$	The position of sentence that contains t_i shown in the description

The terms tend to be important if they are related to the medical domain, thus, we would like to keep a term if it is from a medical related concepts. For instance, the term "*disorder*" is common in the medical domain, so it may not be selected when extracting the Key Terms. However, this term is certainly important if it shows in the phrase "*post-traumatic stress disorder*". We used the **Concept**(t_i) to capture this feature. Specifically, we used MetaMap[1] to identify

---

[1] https://metamap.nlm.nih.gov/.

the medical related concepts from the queries. We will set this feature to true if the term is part of the identified medical concepts.

In addition, the term is also important if it is unique in the medical domain. For example, the word *"vitamin"* is not a common word in the web domain, but it occurs many times in the medical domain. This phenomenon indicates that the terms are more useful in the medical domain. In order to capture this feature, we computed the IDF value of the term in the CDS collection and the one in a regular web domain (i.e., a TREC Web collection). The ratio of these two IDF values is used as the feature. This feature is denoted as **Unique**(t_i).

Abbreviations are widely used in the medical domain, especially to stand for the names of a disease, such as *"PTSD"* and *"UTIs"*, or a diagnostic procedure(*"MRI"*, for example). Correctly locate those medical related abbreviations could improve the retrieval performance. Therefore, we proposed the **Abbr**(t_i) feature to capture this phenomenon. Due to the lack of a comprehensive abbreviation dictionary in medical domain, we used two online dictionaries, i.e., Oxford online dictionary[2] and Merriam-Webster[3], to identify if a term is an abbreviation. This feature will be set to true if the term does not show in any of these two online dictionaries as a English term. Some abbreviations, e.g., *"COLD"*, which stands for *"chronic obstructive lung disease"*, may happen to be a English term, so it can not be captured by the previous feature. We propose to include those terms by using the **All_Cap**(t_i) feature. The feature would be true if the every character in the term is capitalized. Similarly, capitalization could also be an indicator of the domain specific terms or proper nouns. We designed the feature **Capitalized**(t_i) to capture that. This value would be set to true if the term is capitalized.

Ideally, the key concepts can be captured by the nouns and verbs in the sentence. Therefore, we proposed to include the POS tagging as one set of the features. Specifically, there are three features belong to this category, i.e., **Noun**(t_i): whether a term is a noun (or part of the noun phrases), **Verb**(t_i): whether it is a verb (or part of the verb phrases), and **Adj**(t_i): whether it is an adjective.

One straightforward way to measure the effectiveness of the terms is to compute how the retrieval performance will change with and without the term. We adopt the weighted information gain (**wig**(t_i)) from [2] to capture this phenomenon. The **wig**(t_i) is defined as the changes of information from the state of which only average document is retrieved to the state of which the term t_i is actually being used as the query term. It has been shown to be effective in the verbose query processing in web search. We adopt the same equation when computing the wig as in [2], i.e.:

$$wig(t_i) = \frac{\frac{1}{M} \sum_{d \in T_M(t_i)} \log p(t_i|d) - \log p(t_i|\mathcal{C})}{-\log p(t_i|\mathcal{C})} \qquad (1)$$

where the $T_M(t_i)$ is the top returned documents when using term t_i. We set M to 50 in our experiment.

---

[2] https://www.oxforddictionaries.com/.
[3] https://www.merriam-webster.com/.

By observing the shorter version of the EMR and the simplified query in Table 1, we can see that the important information, such as the previous medical history and chief complaint are introduced at the beginning of the EMR. Therefore, we proposed **Rank_des**(t_i) and **Rank_sent**(t_i) to capture the locality information of the terms.

4   Experiments

4.1   Data Sets

In order to perform the machine learning techniques with the proposed features, a data set with importance of the query terms is desired. However, such data set is hard to obtain. We utilized the data sets from TREC Clinical Decision Support track to approximate that, since they contain different versions of the same query. Clinical Decision Support (CDS) track has been held from 2014 to 2016 in TREC. 30 queries are released for each year's task. The example query shown in Table 1 is a query released in 2016. There are three types of queries: The EMR of the patient is the admission note from MIMIC-III, which describes the patient's chief complaint, medical history and other useful information upon admission. This is named as *note* query. The shorter version of EMR and the simplified query are the narratives generated by the organizers based on the *note* query. They are named as *description* and *summary* queries, respectively. By observing the example, it is clear that the *note* query and *description* query are much more verbose than the *summary* query. The *description* query and *summary* query are provided for all three years, while the *note* query is only provided for CDS16. The average query length (number of terms) is reported in Table 3. We can see that even the shortest version of the query, i.e., *summary* query, is still much longer than the verbose query in web domain.

Table 3. Average query length (number of terms) for different query types.

Year	Summary	Description	Note
CDS14	26.97	79.53	–
CDS15	21.87	83.97	–
CDS16	34.4	123.1	248.9

Because the *note* query is not available for all years, we used the *description* and *summary* queries to train the classifier. When generating the training set, for each term occurred in the *description* query, we would consider it as an important term if and only if this term also occurs in the corresponding *summary* query. Term stemming is not used, and the stop words are not removed from the query. The same term shown in different queries is kept because although the term spelling is the same, the feature vectors it generated could be different ($tf_{des}(t_i)$, for instance).

Although the *summary* query is not a strict subset of the *description* query, we argue that the CDS track query set fit this problem setup well for three reasons: First, both the *description* query and the *summary* query are manually created by expert topic developers. Therefore, the quality of the topics is guaranteed. Second, there are 90 queries are given for three years, which generated more than 1000 mapping instances, which is a reasonable size for a training set. Finally, TREC CDS track is a platform for comparing different retrieval models from participants. Therefore, we could compare our performance with the state-of-art runs in this domain.

4.2   Key Terms Identification Results

We tested different machine learning algorithms with the selected features as described in the Sect. 3 using the description queries and summary queries. All 90 queries across three data collections were used, and 5-fold cross validation is applied. The precision, recall and F1 of Key Term identification are reported in the Table 4.

Table 4. Performance of Key Term selection.

	Random forest	Logistic regression	Decision tree	SVM
Precision	0.753	**0.795**	0.642	0.735
Recall	0.631	0.676	**0.821**	0.668
F1	0.686	**0.731**	0.720	0.699

It is clear that Logistic regression and Decision Tree perform better than the other two methods in terms of F1, which indicates that both of these two methods could be useful on identifying the Key Terms. However, since only the Key Terms is what we want to extract from the verbose query, the precision of the identification has a higher priority than recall. Therefore, we chose the logistic regression as our identification models in the following experiments.

4.3   Apply Identified Key Terms for Retrieval

The ultimate goal of this project is to improve the retrieval performance using the selected terms. Therefore, we conducted the experiments using the selected Key Terms. To be specific, we trained the classifier using the *description* query and *summary* query from two data collections, and tested it using the third year's data collection. We did the experiment three times by switching the training data. The results, in terms of infNDCG, are reported in Table 5. We used the Indri with the default Dirichlet smoothing as the retrieval function. The parameter μ is tuned to achieve the best performance.

We included several state-of-art methods as baselines. Noun Phrase is a baseline method described in [17], which only the noun phrases from the description are kept as the query. The Key Concept is a state-of-art baseline method proposed in [2]. Similar as our work, they proposed a set of features to identify the key concepts from the web query. Three features, i.e., $g_t f(c_i)$, $qp(c_i)$ and $qe(c_i)$, are dropped because exterior resources are required to generate these features, while we don't have such access to those resources. Fast QQP is the best query reduction method proposed by Soldaini et al. in [4]. We trained the classifier based on the features described in their paper with the TREC CDS query set, since the query set used in their project is not public available. In addition, we also included the performance of using the *description* query and *summary* query directly, named Description and Summary in the table. The *Summary* baseline could serve as a upper bond for our method since our method is trained against it.

Table 5. Retrieval performance using Key Terms selected from *description* query. The † indicates the improvement over *Description* is statistically significant at 0.05 level based on Wilcoxon signed-rank test.

	CDS14	CDS15	CDS16
Summary	0.1712	0.2067	0.1844
Description	0.1397	0.1615	0.1537
Noun phrase [17]	0.1195	0.1487	0.1322
Key concept [2]	0.1426	0.1657	0.1594
Fast QQP [4]	0.1498	0.1753	0.1584
Learning2extract	**0.1583**†	**0.1779**†	**0.1647**

By comparing the performance in Table 5, we could see that the improvement of our method is significant over the baseline that directly using the *description* query. Our method also outperforms the three baselines, which means that using the domain specific features would be helpful in the problem setup. Note that the Key Concept and Fast QQP methods are not trained as reported in the original paper, so it could be possible that their performances would be improved. The performance of the proposed method is actually close to the results of directly using the summary query. This suggests that our method could successfully identify the useful terms from the verbose query.

In addition to the *description* query, we also tested our method with the *note* query. The *note* query is only available in 2016 data collection. Since only limited training data is available, we used the *description* query from the three year to build the classifier. Figure 1 summarizes the performance of the runs. The performance is reported in terms of infNDCG. The results show that, not surprisingly, the best tuned performance of Key Term selection using the *note* queries is worse than the ones trained on using the description queries for all methods. There could be two reasons, for one thing, the rich information contained in the note query tend to generate more redundant terms which could hurt the performance. For the other thing, the lacking of training example from

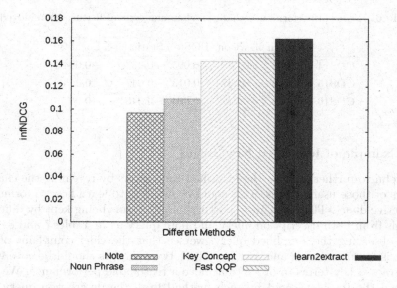

Fig. 1. Retrieval performance using selected terms from *Note* query.

note query to summary query is also a reason for the performance decrease. However, our method could still improve the performance over note query and outperform the other baseline methods, which shows that the proposed feature is robust on identifying Key Terms from verbose queries.

4.4   Feature Importance

In order to better understand the usefulness of the features, we also tested the importance of the each type of features by removing it from the feature set and see how the retrieval performance would change. Specifically, we removed each type of features as described in Table 2, and trained the model with the remaining features. We did the experiments using the description query over the three collections. Table 6 summarize the feature importance analysis results. The negative value in the table means the retrieval performance drops after this type of features is removed from the feature space. Not surprisingly, the domain specific features played an significant role on identifying the Key Terms, while the term statistic features and lexicon features are also promising in important term classification. We further analyzed the performance by removing the features one at a time to see the performance changes. The results indicates that the $abbr(t_i)$, $IDF(t_i)$ and $wig(t_i)$ are the most useful features except the domain features. The POS features did not work well. By further looking into the data, it shows that the noun phrases and verb phrases occur both as the Key Terms and non-Key Terms, so it is hard to learn the pattern from the POS features.

Table 6. Retrieval performance changes when one type of features is removed.

	Domain	Lexicon	POS	Statistical	Locality
CDS14	−0.067	−0.025	−0.005	−0.058	−0.004
CDS15	−0.074	−0.037	0.013	−0.047	0.003
CDS16	−0.066	−0.028	−0.004	−0.045	−0.007

4.5   Example of Identified Key Terms

It is useful to further analyze the identified Key Terms by revealing the characteristics of those useful terms. This could allow users to learn how to formulate an effective query. Therefore, we report the actual terms being kept by different methods from both description query and note query as in Tables 7 and 8.

By observing the simplified query, we see that the chief complaint of the patient is frequent stools and melena. These two concepts should be covered in the extracted key terms in order to achieve a reasonable performance. We first examined the terms selected by each method from the description query (i.e. Table 7). For the Noun Phrase method, although both two concepts are covered in the shorter version, too many irrelevant terms have been kept since they are nouns. Thus, the identified query is drifted because of these noisy terms. The Key Concept and Fast QQP methods solved this problem to a certain extent by involving other features when selecting the query terms, but they still suffer from the noisy terms, such as "nursing home" and "approximately". In addition,

Table 7. Identified Key Terms from the *description* query

Methods	Identified key terms
A shorter version of the EMR	78 M transferred to nursing home for rehab after CABG. Reportedly readmitted with a small NQWMI. Yesterday, he was noted to have a melanotic stool and then today he had approximately 9 loose BM w/ some melena and some frank blood just prior to transfer, unclear quantity
Simplified query	A 78 year old male presents with frequent stools and melena
Noun phrase	Nursing home a small NQWMI a melanotic stool 9 loose BM some melena and some frank blood
Key concept	Nursing home CABG a small NQWMI noted stool prior to transfer
Fast QQP	Nursing CABG readmitted with a small NQWMI melanotic stool approximately loose melena
Learn2extract	Rehab CABG NQWMI melanotic stool BM melena frank blood

Table 8. Identified Key Terms from the *note* query

Query type	Query content
The EMR of the patient	78 M w/ pmh of CABG in early [**Month (only) 3**] at [**Hospital6 4406**] (transferred to nursing home for rehab on [**12-8**] after several falls out of bed.) He was then readmitted to [**Hospital6 1749**] on [**3120-12-11**] after developing acute pulmonary edema/CHF/unresponsiveness?. There was a question whether he had a small MI; he reportedly had a small NQWMI. He improved with diuresis and was not intubated. Yesterday, he was noted to have a melanotic stool earlier this evening and then approximately 9 loose BM w/ some melena and some frank blood just prior to transfer, unclear quantity
Simplified query	A 78 year old male presents with frequent stools and melena
Noun phrase	CABG home acute pulmonary edema unresponsiveness a small MI NQWMI diuresis loose BM melanotic stool frank blood unclear quantity
Key concept	CABG nursing home acute pulmonary edema CHF unresponsiveness small NQWMI melanotic stool loose BM
Fast QQP	pmh CABG nursing home falls bed pulmonary edema CHF unresponsiveness diuresis was not intubated melanotic loose frank blood
Learn2extract	pmh CABG nursing home rehab pulmonary edema CHF NQWMI melanotic stool loose BM melena frank blood

they missed some important terms in the shorter version (such as "melena" for Key Concept). This would also hurt the performance too. Our method, on the other hand, successfully identified these two aspects, and our method could also bring additional useful term (i.e., "frank blood") to the query.

We then examined the key term selection from the note query as shown in Table 8. Since the note query contains more information than the description query, every methods generated a much longer key terms list comparing with the key terms selected based on description query. This also explains the performance decrease of using note query as shown in Table 1. After a close look at the identified key terms in Table 8, we find that although all methods are suffered from the query drifting problem because of the additional terms, our method contains the least, yet most useful, terms comparing with the other methods.

5   Conclusion

In this work, we proposed a new set of features to identify the Key Terms from verbose query for retrieval in medical domain. Experiment results over three data collections show that using the selected Key Terms could significantly improve the retrieval performance than directly using the original verbose query, and it also outperform two strong baselines.

There are many directions that we plan to work on as future work. First, we would like to explore more features, especially more domain features, to enrich the feature space. Second, instead of the classifier, we would like to design a weighting schema for each term based on their importance. Finally, it would be interesting to see how the proposed feature would work with the other machine learning algorithm, such as CNN, to solve this problem.

Acknowledgments. This research was supported by the U.S. National Science Foundation under IIS-1423002.

References

1. Gupta, M., Bendersky, M.: Information retrieval with verbose queries. In: Proceedings of the 38th International ACM SIGIR Conference on Research and Development in Information Retrieval (SIGIR 2015), New York, NY, USA, pp. 1121–1124. ACM (2015)
2. Bendersky, M., Croft, W.B.: Discovering key concepts in verbose queries. In: Proceedings of the 31st Annual International ACM SIGIR Conference on Research and Development in Information Retrieval (SIGIR 2008), pp. 491–498 (2008)
3. Jones, R., Fain, D.C.: Query word deletion prediction. In: Proceedings of the 26th Annual International ACM SIGIR Conference on Research and Development in Information Retrieval (SIGIR 2003), New York, NY, USA, pp. 435–436. ACM (2003)
4. Soldaini, L., Cohan, A., Yates, A., Goharian, N., Frieder, O.: Retrieving medical literature for clinical decision support. In: Hanbury, A., Kazai, G., Rauber, A., Fuhr, N. (eds.) ECIR 2015. LNCS, vol. 9022, pp. 538–549. Springer, Cham (2015). doi:10.1007/978-3-319-16354-3_59
5. Kumaran, G., Carvalho, V.R.: Reducing long queries using query quality predictors. In: Proceedings of the 32nd International ACM SIGIR Conference on Research and Development in Information Retrieval (SIGIR 2009), pp. 564–571 (2009)
6. Turney, P.D.: Learning algorithms for keyphrase extraction. Inf. Retr. **2**(4), 303–336 (2000)
7. Díaz-Galiano, M.C., Martín-Valdivia, M., Ureña López, L.A.: Query expansion with a medical ontology to improve a multimodal information retrieval system. Comput. Biol. Med. **39**(4), 396–403 (2009)
8. Martinez, D., Otegi, A., Soroa, A., Agirre, E.: Improving search over electronic health records using umls-based query expansion through random walks. J. Biomed. Inform. **51**, 100–106 (2014)
9. Yang, C., He, B., Xu, J.: Integrating feedback-based semantic evidence to enhance retrieval effectiveness for clinical decision support. In: Chen, L., Jensen, C.S., Shahabi, C., Yang, X., Lian, X. (eds.) APWeb-WAIM 2017. LNCS, vol. 10367, pp. 153–168. Springer, Cham (2017). doi:10.1007/978-3-319-63564-4_13
10. Zhu, D., Carterette, B.: Combining multi-level evidence for medical record retrieval. In: Proceedings of the 2012 International Workshop on Smart Health and Wellbeing (SHB 2012), pp. 49–56 (2012)

11. Limsopatham, N., Macdonald, C., Ounis, I.: Aggregating evidence from hospital departments to improve medical records search. In: Serdyukov, P., et al. (eds.) ECIR 2013. LNCS, vol. 7814, pp. 279–291. Springer, Heidelberg (2013). doi:10.1007/978-3-642-36973-5_24
12. Wang, Y., Fang, H.: Exploring the query expansion methods for concept based representation. In: TREC 2014 (2014)
13. Limsopatham, N., Macdonald, C., Ounis, I.: Learning to combine representations for medical records search. In: Proceedings of SIGIR 2013 (2013)
14. Qi, Y., Laquerre, P.F.: Retrieving medical records: NEC Labs America at TREC 2012 medical record track. In: TREC 2012 (2012)
15. Koopman, B., Zuccon, G., Nguyen, A., Vickers, D., Butt, L., Bruza, P.D.: Exploiting SNOMED CT concepts & relationships for clinical information retrieval: Australian e-health research centre and Queensland University of Technology at the TREC 2012 medical track. In: TREC 2012 (2012)
16. Wang, Y., Liu, X., Fang, H.: A study of concept-based weighting regularization for medical records search. In: ACL 2014 (2014)
17. Wang, Y., Fang, H.: Extracting useful information from clinical notes. In: TREC 2016 (2016)

FORK: Feedback-Aware ObjectRank-Based Keyword Search over Linked Data

Takahiro Komamizu[✉], Sayami Okumura, Toshiyuki Amagasa,
and Hiroyuki Kitagawa

University of Tsukuba, Tsukuba, Japan
taka-coma@acm.org, okumura@kde.cs.tsukuba.ac.jp,
{amagasa,kitagawa}@cs.tsukuba.ac.jp

Abstract. Ranking quality for keyword search over Linked Data (LD) is crucial when users look for entities from LD, since datasets in LD have complicated structures as well as much contents. This paper proposes a keyword search method, FORK, which ranks entities in LD by ObjectRank, a well-known link-structure analysis algorithm that can deal with different types of nodes and edges. The first attempt of applying ObjectRank to LD search reveals that ObjectRank with inappropriate settings gives worse ranking results than PageRank which is equivalent to ObjectRank with all the same authority transfer weights. Therefore, deriving appropriate authority transfer weights is the most important issue for encouraging ObjectRank in LD search. FORK involves a relevance feedback algorithm to modify the authority transfer weights according with users' relevance judgements for ranking results. The experimental evaluation of ranking qualities using an entity search benchmark showcases the effectiveness of FORK, and it proves ObjectRank is more feasible raking method for LD search than PageRank and other comparative baselines including information retrieval techniques and graph analytic methods.

Keywords: Keyword Search over Linked Data · ObjectRank-based ranking · Relevance feedback · Authority transfer graph modification

1 Introduction

Linked Data (or LD) is an emerging data publishing movement triggered by Sir Tim Berners-Lee [7]. LD datasets contain diverse sorts of real-world facts and are connected through the Web. For instance, DBpedia[1] contains extracted resources from Wikipedia and is one of the most popular LD datasets acting as the hub of various LD datasets; GeoNames[2] includes world-wide geographical data; and scholarly digital libraries also publish their datasets as LD datasets (e.g., *LOD for Conferences in Computer Science* by Springer[3] and D2R server for DBLP[4]).

[1] http://wiki.dbpedia.org/.
[2] http://www.geonames.org/.
[3] http://lod.springer.com/wiki/bin/view/Linked+Open+Data/About.
[4] http://dblp.l3s.de/d2r/snorql/.

© Springer International Publishing AG 2017
W.-K. Sung et al. (Eds.): AIRS 2017, LNCS 10648, pp. 58–70, 2017.
https://doi.org/10.1007/978-3-319-70145-5_5

Keyword search [5] is a promising technique to facilitate querying over LD but it suffers from the ranking difficulty. In contrast with traditional graph data search (e.g., Web search) which have single *types* of vertices (like *Web pages* in Web search), LD has a large variety of vertices specified by rdf:type. PageRank [16] which is successfully applied to the traditional graph search is not feasible enough for LD search because of the variety. Therefore, this paper applies ObjectRank [4] which has been developed with close idea for PageRank but can handle multiple types of objects composing graph data.

As justified in many researches (like [10]), a research issue arises when applying ObjectRank to keyword search on LD, that is, how to design authority transfer graphs (i.e., schema and data graphs) from LD (Sect. 2). In many situations, authority transfer graphs are manually designed over different types of objects, however, LD datasets tend to have large number of types. The experiment in this paper shows an interesting fact that ObjectRank with randomly assigned authority transfer weights is (on average) lesser ranking quality than PageRank. This fact indicates that appropriate authority transfer weights are highly desirable to test the feasibility of ObjectRank on LD search.

This paper employs a query re-evaluation technique called **relevance feedback** [3] to learn authority transfer weights w.r.t. users' experiences for search results (Sect. 3). In relevance feedback, users provide relevance judgements on search results (typically top-k results), and search systems re-evaluate search results to make the relevant results as well as similar results to the relevant results higher ranking. Traditional relevance feedback arranges importances of attributes of objects, while relevance feedback for ObjectRank handle arrangements of authority transfer weights, and this is not straightforward. This paper proposes a relevance feedback-aware keyword search, **FORK** (**F**eedback-aware **O**bject**R**ank-based **K**eyword Search over Linked Data).

This paper demonstrates effectiveness of FORK with the real-world, DBpedia, and an entity search benchmark [6] on DBpedia (Sect. 4). The preliminary experiment in this paper attempts to find relationships between the number of relevance feedbacks and ranking qualities. The experiment reveals that the more relevance feedbacks the higher ranking quality, and ranking qualities can be saturated when the enough number of relevance feedbacks are given. The main experiment compares the ranking quality of FORK with that of baseline approaches, and showcases that FORK achieves the bast ranking method among them. Therefore, this paper concludes that FORK successfully adjusts and derives appropriate authority transfer weights to improve rankings of LD search.

The major contribution of this paper is **FORK**, and the following contributions support FORK in this paper.

- **Ranking Quality**: This paper ensures that ObjectRank with appropriate authority transfer graphs improves the state-of-the-art in keyword search over LD. The existing work rely either on topological or textual similarities, however, ObjectRank can handle both in a simple scheme.

- **Appropriate Weight Learning Algorithm**: FORK involves an authority transfer weight learning algorithm inspired from relevance feedback techniques. Based on relevance judgements from users for results, the algorithm makes the authority transfer weights closer to the users' expectations.

2　ObjectRank: Preliminary

ObjectRank [4] requires two graph data, namely, authority transfer schema graph (or schema graph) and authority transfer data graph (or data graph). In the ObjectRank setting, a dataset contains set C of data types (or classes) of objects. Schema graph models relationships between classes. On schema graph, the vertices represent classes, the edges represent directed relationships between classes, and each edge is weighted for transferring authority from source to destination.

A data graph is derived from the schema graph by mapping weights onto graphs consisting of objects of classes and relationships between objects. For an edge e which source vertex $src \in O_c$ (where O_x denotes a set of objects of class x) belongs to class c and destination vertex $dst \in O_d$ belongs to class d, the weight $w_{src,dst}$ of edge (src, dst) is calculated as follows:

$$w_{src,dst} = \frac{W_S((c,d))}{outdeg(src,d)} \tag{1}$$

where $W_S((c,d))$ returns the weight of an edge between classes c and d in schema graph, and $outdeg(src, d)$ function returns the number of outgoing edges from src and destination vertices of the edges are of class d.

From a data graph, authority scores of objects are calculated as analogous to PageRank. In ObjectRank, two sorts of authorities (i.e., global ObjectRank and query-specific ObjectRank) are computed in order to evaluate both topological authorities and query-centric authorities.

$$\mathbf{r}_g^{(t+1)} = dA\mathbf{r}_g^{(t)} + \frac{1-d}{|O|}\mathbf{e} \tag{2}$$

$$\mathbf{r}_q^{(t+1)} = dA\mathbf{r}_q^{(t)} + \frac{1-d}{|S(q)|}\mathbf{s} \tag{3}$$

Global ObjectRank is iteratively calculated as Eq. 2, where $\mathbf{r}_g^{(t)}$ is an authority vector of objects in t-th iteration, A is a $|O| \times |O|$ adjacency matrix composing of weights between objects in data graph, d is a dumping factor balancing authority transitivity in the graph and random-jump, and $\mathbf{e}$ is a vector of all 1 with length $|O|$. Query-specific ObjectRank is iteratively calculated as Eq. 3, where $\mathbf{r}_q^{(t)}$ is an authority vector of objects in t-th iteration, $S(q)$ is a set of vertices which meet with query q, and $\mathbf{s}$ is a vector of 1 for vertices matching with q, 0 otherwise. Overall ObjectRank score $\mathbf{r}$ for given query q is weighted combination of the global and query-specific authorities. The combination is formulated as Eq. 4

where u is an adjustable parameter of effects of query-specific authorities, and ∘ represents Hadamard product (or element-wise multiplication).

$$\mathbf{r} = \mathbf{r}_g \circ (\mathbf{r}_q)^u \tag{4}$$

Formally, in consequence, ObjectRank computation is defined as Definition 1.

Definition 1 (ObjectRank computation). *Given data graph DG and a keyword query q, ObjectRank values for q on DG are calculated Eq. 4 by combination of global and query-specific ObjectRank computations as in Eqs. 2 and 3 based on an adjacency matrix A derived from DG.* □

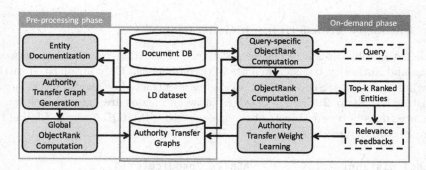

Fig. 1. Overview of FORK.

3   FORK: Proposed Method

Figure 1 draws an overview of FORK, where white-colored objects are given in advance (i.e., datasets and storages) or by users (queries and feedbacks) and gray-colored objects are processes. The procedure of FORK can be divided into two phases, namely, pre-processing phase and on-demand phase. The former phase processes given LD datasets to prepare for ranking entities by ObjectRank scores. The pre-processing phase includes three processes, namely, *entity documentization, authority transfer graph generation*, and *global ObjectRank computation*. The latter phase computes rankings of entities for given user queries and modifies authority transfer graphs based on users' relevance judgements for the ranking results. The on-demand phase consists of three processes, namely, *query-specific ObjectRank computation, ObjectRank computation*, and *authority transfer weight learning*. The subsequent sections introduce entity documentization in Sect. 3.1, authority transfer graph generation in Sect. 3.2, and authority transfer weight learning in Sect. 3.3.

3.1　Entity Documentization

This paper employs a reasonable documentization technique by Sinha et al. [19]. They represent an object by concatenation of surrounding attributes and their values as a document. This paper applies this idea for LD datasets to documentize entities. The following SPARQL query provides a document of an entity (by specifying on the placeholder "<entity>").

```
SELECT ?value
WHERE { <entity> ?predicate ?value.
     FILTER(isLiteral(?value)) }
```
Listing 1.1. SPARQL query for documentization.

3.2　Authority Transfer Graph Generation

A schema graph consists of classes (or types) of entities, relationship between classes and authority transfer weights between classes. They are derived in the following three steps. **Step 1**: Set C of classes is extracted by the SPARQL query in Listing 1.2. **Step 2**: Set E_S of edges between classes are extracted as far as there are edges between objects of the classes with SPARQL query in Listing 1.3 where c1 and c2 are classes in C.

```
SELECT distinct ?class
WHERE{?s rdf:type ?class}
```
Listing 1.2. Listing classes.

```
ASK{?s ?predicate ?d.
   ?s rdf:type <c1>. ?d rdf:type <c2>.}
```
Listing 1.3. Presence of edges.

Step 3: Authority transfer weights W_S on schema graph are initialized with random values, due to the absence of proper criteria to set the weights.

A data graph is composed of objects (or entities), and authority transfer weights between entities. Entities and edges corresponding with classes are extracted by Listing 1.4 where c1 and c2 are connected classes on the schema graph. As a result of the query, the variables ?s and ?d in the query gives set O of objects and set E_O of edges between objects, simultaneously.

```
SELECT distinct ?s ?d
WHERE{?s ?predicate ?d. ?s rdf:type <c1>. ?d rdf:type <c2>.}
```
Listing 1.4. Objects corresponding with schema graph.

As Eq. 1, the weights W_D on data graph are derived from schema graph based on degrees of each object.

3.3　Authority Transfer Weight Learning

Link-structure analysis algorithms give higher scores to vertices which random surfers arrive with high probability. A vertex having many in-coming edges with high authority transfer weights receives large amount of authorities from other vertices, therefore it obtains high score. On the other hand, a vertex having many out-going edges with high authority transfer weights supplies most of its authority to the connected vertices, therefore, it obtains low score.

The key idea in this paper is classes of objects with positive feedbacks should receive more authority and leak less authority, while those with negative feedbacks should supply more authority and receive less authority. The aforementioned facts indicate that if a vertex should obtain high score, it should have in-coming edges with high authority transfer weights and out-going edges with low authority transfer weights. While, if a vertex should obtain low score, it should have in-coming edges with low authority transfer weights and out-going edges with high authority transfer weights. Figure 2 overviews individual mechanisms for modifying weights on schema graph. The above figures represent an edge between classes, $C1$ (source vertex) and $C2$ (destination vertex), with weight $w_{C1,C2}$. By feedbacks on $C1$ or $C2$, modifications of weights are classified into four types. (a) given positive feedback to $C1$, in order to avoid leak-out of authority from $C1$, $w_{C1,C2}$ is decreased, while (b) if the feedback on $C1$ is negative, $w_{C1,C2}$ is increased. On the contrary to $C1$, (c) given positive feedback to $C2$, to increase the flowing-in authority, $w_{C1,C2}$ is increased, on the other hand, (d) negative feedback for $C2$ decreases $w_{C1,C2}$.

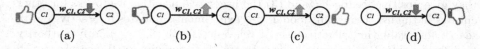

(a)   (b)   (c)   (d)

Fig. 2. Four situations of authority transfer weight modification on schema graph.

This paper formulates the aforementioned idea to compute the shift R_{ij} of authority transfer weights as follows:

$$R_{ij} = \left(1 - \sum_{k=1}^{NP_i} \frac{1}{\beta^k}\right)\left(1 + \sum_{k=1}^{NN_i} \frac{1}{\alpha^k}\right)\left(1 + \sum_{k=1}^{NP_j} \frac{1}{\alpha^k}\right)\left(1 - \sum_{k=1}^{NN_j} \frac{1}{\beta^k}\right) \quad (5)$$

where NP_i and NN_i are the numbers of relevant and non-relevant objects of class C_i, and α and β are adjustable parameters for increments and decrements of the edge weight. The order of bracketed terms is corresponding with Fig. 2. The shift updates weight $W_S(C_i, C_j)$ in the schema graph as $W_S(C_i, C_j) = W_S(C_i, C_j) \cdot R_{ij}$. Based on the updated authority transfer weights of schema graph, FORK calculates data graph as well as global ObjectRank values.

4   Experimental Evaluation

This paper proves the efficiency of FORK by answering the following questions.

Q1: *Does FORK improve ranking quality of ObjectRank by adjusting authority transfer weights according with users' relevance judgements?*

Q2: *Is ObjectRank-based ranking method a better ranking method comparing with other ranking methods like IR-based and graph analysis-based methods?*

4.1   Experimental Settings

This experiment evaluates FORK by one of the largest and the most popular dataset DBpedia 3.9[5] and test collection [6] which is a benchmark dataset of entity search task. Though the test collection containing 485 queries for entity search, 61 queries are chosen for testing FORK because some queries are too complicated (i.e., more than five words which cannot be answered by some baselines) and are represented in natural languages which FORK cannot handle. The test collection includes answer entities matching with each query.

To emulate relevance feedbacks on each query in the test collection, this paper employs **honest-and-static user model**. In the honest-and-static model, users behave that (1) all results in the ranked list are evaluated, (2) wherever the users find true entities (in the answer set of the test collection) in the list, they give positive feedbacks, and (3) the users give negative feedbacks for other entities in the list. Using this model, this experiment simulates interactive entity search for evaluating effectiveness of relevance feedbacks to modify authority transfer weights. Ranking quality is evaluated by **precision@10** measure, which is calculated as a ratio of true results in the top-10 ranked results. Parameter settings for FORK are $k = 10$ and $\alpha = \beta = 2$. k is defined based on evaluation strategy (i.e., precision@10). α and β are simply chosen from experiences.

4.2   Ranking Quality over the Number of Feedbacks

This section is designed to answer **Q1** by observing ranking accuracy (i.e., precision@10) over the relevance feedbacks. Expectation for the evaluation is that ranking qualities gradually increase as the number of feedbacks increases, and it is thus possible to say FORK gives proper authority transfer weights according with users' relevance feedbacks.

Figure 3(a) showcases ranking quality improvements over five relevance feedbacks, and it indicates that the authority transfer weight learning algorithm in FORK learns the weights properly from users' relevance feedbacks. Each bar in the figure indicates precision@10 scores for the authority transfer weights at the moment when the specific number of relevance feedbacks are given. The figure indicates that the ranking quality increases as the number of relevance feedbacks

---

[5] http://wiki.dbpedia.org/services-resources/datasets/data-set-39.

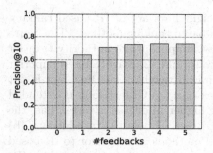 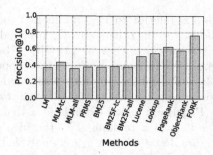

(a) Precision@10 over #feedbacks of ObjectRank-based ranking with FORK.

(b) Precision@10 for the ranking methods.

Fig. 3. Evaluation results.

increases. The first three relevance feedbacks increase ranking quality more drastically than the latter feedbacks. Thus, large number of relevance feedbacks do not much contribute to the ranking quality.

Consequently, the following gives the answer for **Q1**.

A1: *FORK improves the ranking quality of ObjectRank by modifying authority transfer weights. This experimentally proves that FORK provides appropriate authority transfer weights reflecting users' search experiences. However, there is still space left to improve the ranking quality due to the saturation of ranking qualities.*

4.3   Ranking Effectiveness

This section is designed to answer **Q2** by comparing ObjectRank-based ranking methods with other comparative methods including IR-based and graph analysis-based ranking methods. The comparative methods are very static, meaning that rankings from their results do not change if feedbacks are given. Therefore, for fairness of comparison, this experiment includes ObjectRank without relevance feedback (i.e., randomly assigned authority transfer weights) and ObjectRank with learned weights by FORK. Expectation of this experiment is that ObjectRank with appropriate authority transfer weights outperforms others.

This experiment compares the following ranking methods (texts in parentheses represent shorter names used in the following discussion). **IR-based ranking**: language model (LM) [21], mixture of language model (MLM-tc) [15], probabilistic retrieval model for semi-structured data (PRMS) [12], BM25 [18], fielded extensions [14] of MLM and BM25, (namely, MLM-all, BM25F-tc, BM25F-all), and Apache Lucene [1] (Lucene). **Graph analysis-based ranking**: DBpedia Lookup [2] (Lookup) which is based on degrees of vertices, PageRank [16], and ObjectRank with randomly initialized authority transfer weights (ObjectRank).

Figure 3(b) shows comparative results of precision@10, and it indicates that ObjectRank with FORK outperforms others and ObjectRank with random weights is inferior to PageRank. Left-most eight bars are of IR-based ranking methods, and the others are of graph analysis-based ranking methods. The figure indicates two facts that are (1) ObjectRank with appropriate authority transfer weights (FORK) outperforms other approaches, and (2) PageRank is superior to ObjectRank-based ranking with random initial weights. The first fact is not big surprising because of the following two reasons: (a) ObjectRank is extended from PageRank which extends idea that the higher in-degree a vertex has, the higher it is in a ranking (i.e., the idea of DBpedia Lookup). In addition, (b) the figure describes that graph analysis-based methods are superior to IR-based ones. This is because these methods can access to more information than IR-based ones. The second fact is a little surprising that PageRank outperforms ObjectRank with random weights. PageRank can be seen as a special case of ObjectRank, that is, PageRank is equivalent to ObjectRank if all the authority transfer weights in ObjectRank are equal. Therefore, this experiment concludes that authority transfer weights should be carefully designed.

Consequently, the following give the answer for **Q2**.

A2: *ObjectRank is the best ranking method if the authority transfer weights are appropriately designed. An interesting fact is that randomly initialized weights produce (on average) worse ranking quality than equal weights for all edges (i.e., PageRank).*

4.4   Ranking Showcase

This section demonstrates the changes of ranking during users' interactions by breaking down rankings during relevance feedbacks one by one. Herein, the showcase include an example query "harry potter" in the test collection which expects to entities related with movies and characters of Harry Potter. Table 1 displays a sequence of ranking results over relevance feedbacks starting from initial ranking until five feedbacks.

The series of tables indicate that ranking quality gradually increases as the number of feedbacks increases. In this example, the initial ranking result in Table 1(a) includes seven true results but non-relevant results are in higher ranking. For example, an entity "Harry Potter" is obviously ranked on the top, however, the second entity "Harry Potter Fan Zone" is an Australian Harry Potter fan-site, that is not relevant to the query but ranked higher. After first relevance feedback on the results (Table 1(b)), the ranking quality is improved. For instance, the entity "Harry Potter Fan Zone" gets two ranks lower. However, results still include three non-relevant entities. More and more feedbacks are provided (Table 1(c) to (f)), relevant entities get involved in the ranking and relevant entities get higher position in the ranking. Finally, "Harry Potter Fan Zone" gets sixth rank in the list and only two non-relevant entities (i.e., "Harry Potter Fan Zone" and "Lego Harry Potter") are included. As Sect. 4.2 indicates, the last two rankings (i.e., Table 1(e) and (f)) are same.

Table 1. Rankings through relevance feedbacks on query "harry potter".

(a) Ranking results w/o feedback.

Rank	Entity Label	Judg.
1	Harry Potter	✓
2	Harry Potter Fan Zone	
3	Harry Potter (character)	✓
4	List of Harry Potter characters	✓
5	Harry Potter and the Prisoner ...	✓
6	Harper Marshall	
7	Lego Harry Potter	
8	Harry Potter and the Deathly ...	✓
9	J. K. Rowling	✓
10	Magical creatures in Harry ...	✓

(b) Ranking results w/ first feedback.

Rank	Entity Label	Judg.
1	Harry Potter	✓
2	Harry Potter (character)	✓
3	List of Harry Potter characters	✓
4	Harry Potter Fan Zone	
5	Harry Potter and the Prisoner ...	✓
6	Harry Potter and the Deathly ...	✓
7	Lego Harry Potter	
8	Magical creatures in Harry ...	✓
9	J. K. Rowling	✓
10	Harper Marshall	

(c) Ranking results w/ second feedback.

Rank	Entity Label	Judg.
1	Harry Potter	✓
2	Harry Potter (character)	✓
3	List of Harry Potter characters	✓
4	Harry Potter Fan Zone	
5	Harry Potter and the Prisoner ...	✓
6	Harry Potter and the Deathly ...	✓
7	Magical creatures in Harry ...	✓
8	Lego Harry Potter	
9	J. K. Rowling	✓
10	Harry Potter and the Goblet ...	✓

(d) Ranking results w/ third feedback.

Rank	Entity Label	Judg.
1	Harry Potter	✓
2	Harry Potter (character)	✓
3	List of Harry Potter characters	✓
4	Harry Potter and the Prisoner ...	✓
5	Harry Potter Fan Zone	
6	Harry Potter and the Deathly ...	✓
7	Magical creatures in Harry ...	✓
8	Lego Harry Potter	
9	J. K. Rowling	✓
10	Harry Potter and the Goblet ...	✓

(e) Ranking results w/ fourth feedback.

Rank	Entity Label	Judg.
1	Harry Potter	✓
2	Harry Potter (character)	✓
3	List of Harry Potter characters	✓
4	Harry Potter and the Prisoner ...	✓
5	Harry Potter and the Deathly ...	✓
6	Harry Potter Fan Zone	
7	Magical creatures in Harry ...	✓
8	J. K. Rowling	✓
9	Lego Harry Potter	
10	Harry Potter and the Goblet ...	✓

(f) Ranking results w/ fifth feedback.

Rank	Entity Label	Judg.
1	Harry Potter	✓
2	Harry Potter (character)	✓
3	List of Harry Potter characters	✓
4	Harry Potter and the Prisoner ...	✓
5	Harry Potter and the Deathly ...	✓
6	Harry Potter Fan Zone	
7	Magical creatures in Harry ...	✓
8	J. K. Rowling	✓
9	Lego Harry Potter	
10	Harry Potter and the Goblet ...	✓

5   Related Work

Exploiting ObjectRank is a straightforward idea but there is a difficulty to design proper authority transfer graphs. As justified in many researches (like [10]), this difficulty has been known, but, to the best of our knowledge, this paper is the first attempt to tackle this difficulty. Therefore, existing approaches have employed different ranking methods like IR-based approaches and graph analysis-based approaches. The following sections discuss entity ranking in Sect. 5.1, ObjectRank extension in Sect. 5.2 and keyword search in Sect. 5.3.

5.1   Entity Ranking

Ranking strategies are roughly classified into (1) IR-oriented ranking and (2) graph structure-based ranking. The former attempt to apply traditional ranking

methodologies in IR researches to rank entities in LD datasets. They regard entities as documents by employing surrounding textual information of the entities.
The latter utilize topological information of LD datasets to derive importances
of entities. They are inspired from graph analysis researches. Therefore, the following discuss both ranking methodologies as related work.

IR-Oriented Ranking. Pound et al. [17] and Balog et al. [5] have respectively
proposed a TF-IDF based ranking method and a language model-based ranking method. These methods are well-studied techniques in IR communities and
they are novel to apply these techniques to entity search on LD datasets. Their
approaches first documentize entities as this paper, second vectorize the docu-
mentized entities by each methodology (i.e., TF-IDF and language modeling),
then compare with vectorized queries, and rank results based on ranking methods on the vector space model. As the experiment examines, these techniques
are inferior to graph structure-based ranking methods including PageRank and
ObjectRank (by FORK).

Graph Analysis-Based Ranking. PageRank [16] is a mature technique and
several researches (e.g., [8,9]) extend it for entity search. Swoogle [8] is a search
and metadata engine for the semantic web which crawls, indexes, and ranks
semantic web documents based on PageRank algorithm. Ding et al. [9] have
proposed ranking algorithm called OntoRank which also based on PageRank
algorithm. Even OntoRank takes graph structure of LD datasets into account,
their model still considers relationships between entities equally. ObjectRank is
more flexible in nature, thus, in this point, FORK is more advantageous.

5.2   ObjectRank Extension

Even though many researches justifying the weak point of ObjectRank, that is,
manual designing authority transfer graphs is bothersome and almost impossible,
there have been no research dealing with the point. Slightly related work is proposed by Varadarajan et al. [20]. They have proposed ObjectRank2, a relevance
feedback approach to improve users' search experiences. They have introduced
vector space model for query-specific ObjectRank vector (i.e., $\mathbf{s}$ in Eq. 3), and it
becomes possible to employ relevance feedback-based query re-evaluation techniques in the context of ObjectRank-based ranking. However, their approach
cannot obtain appropriate authority transfer graphs.

5.3   Keyword Search over Linked Data

Keyword search on LD datasets is attempted by various viewpoints. From a
keyword search over graph data point of view, Le et al. [13] have proposed a
scalable keyword search which adopt the idea of BLINKS [11] (which is keyword
search for node-labeled graph data) to keyword search on LD datasets. They
attempt to find smallest subgraphs which contains all query keywords in vertices
of LD datasets. In this point of view, targets of keyword search are not limited,

thus documentization is not necessary, while it requires to understand whether relationships (or predicates) in the result subgraphs are reasonable.

In another point of view, structures of LD datasets are taken into consideration. Sinha et al. [19] have proposed a simple vector space model-based keyword search. They vectorize entities by terms in literals of surrounding predicates. Based on the vectorization, they can apply traditional IR-oriented matching and ranking mechanisms to find and rank entities. Pound et al. [17] have proposed query type-based keyword search interface, which, at first, classifies queries into query types such as entity search, entity type search, entity attribute search, and relation search. Based on the query types, they execute different query processing strategies to obtain results.

6   Conclusion

This paper deals with keyword search over Linked Data by applying entity documentization technique and ranking mechanism using ObjectRank. The issue for ObjectRank that designing appropriate authority transfer graphs is solved by relevance feedback-based authority transfer graph modification method. The whole process is composed in FORK, the proposed system. Experimental evaluations demonstrate effectiveness of ObjectRank with modified authority graphs by FORK. The evaluations reveal that graph structure-based ranking methods are superior to IR-oriented ranking methods, in consequence, graph topological information are more important to evaluate entities in LD datasets.

Acknowledgement. This research was partly supported by the program *Research and Development on Real World Big Data Integration and Analysis* of RIKEN, Japan, and Fujitsu Laboratory, *APE29707.*

References

1. Apache Lucene. https://lucene.apache.org/
2. DBpedia Lookup. https://github.com/dbpedia/lookup
3. Baeza-Yates, R., Ribeiro-Neto, B., et al.: Modern Information Retrieval, vol. 463. ACM Press, New York (1999)
4. Balmin, A., Hristidis, V., Papakonstantinou, Y.: ObjectRank: authority-based keyword search in databases. In: VLDB 2004, pp. 564–575 (2004)
5. Balog, K., Bron, M., de Rijke, M.: Query modeling for entity search based on terms, categories, and examples. ACM Trans. Inf. Syst. **29**(4), 22:1–22:31 (2011)
6. Balog, K., Neumayer, R.: A test collection for entity search in DBpedia. In: SIGIR 2013, pp. 737–740 (2013)
7. Berners-Lee, T.: Linked data. https://www.w3.org/DesignIssues/LinkedData.html
8. Ding, L., Finin, T.W., Joshi, A., Pan, R., Cost, R.S., Peng, Y., Reddivari, P., Doshi, V., Sachs, J.: Swoogle: a search and metadata engine for the semantic web. In: CIKM 2004, pp. 652–659 (2004)
9. Ding, L., Pan, R., Finin, T., Joshi, A., Peng, Y., Kolari, P.: Finding and ranking knowledge on the semantic web. In: Gil, Y., Motta, E., Benjamins, V.R., Musen, M.A. (eds.) ISWC 2005. LNCS, vol. 3729, pp. 156–170. Springer, Heidelberg (2005). doi:10.1007/11574620_14

10. Harth, A., Kinsella, S., Decker, S.: Using naming authority to rank data and ontologies for web search. In: Bernstein, A., Karger, D.R., Heath, T., Feigenbaum, L., Maynard, D., Motta, E., Thirunarayan, K. (eds.) ISWC 2009. LNCS, vol. 5823, pp. 277–292. Springer, Heidelberg (2009). doi:10.1007/978-3-642-04930-9_18

11. He, H., Wang, H., Yang, J., Yu, P.S.: BLINKS: ranked keyword searches on graphs. In: SIGMOD 2007, pp. 305–316 (2007)

12. Kim, J., Xue, X., Croft, W.B.: A probabilistic retrieval model for semistructured data. In: Boughanem, M., Berrut, C., Mothe, J., Soule-Dupuy, C. (eds.) ECIR 2009. LNCS, vol. 5478, pp. 228–239. Springer, Heidelberg (2009). doi:10.1007/978-3-642-00958-7_22

13. Le, W., Li, F., Kementsietsidis, A., Duan, S.: Scalable keyword search on large RDF data. IEEE Trans. Knowl. Data Eng. 26(11), 2774–2788 (2014)

14. Neumayer, R., Balog, K., Nørvåg, K.: When simple is (more than) good enough: effective semantic search with (almost) no semantics. In: Baeza-Yates, R., Vries, A.P., Zaragoza, H., Cambazoglu, B.B., Murdock, V., Lempel, R., Silvestri, F. (eds.) ECIR 2012. LNCS, vol. 7224, pp. 540–543. Springer, Heidelberg (2012). doi:10.1007/978-3-642-28997-2_59

15. Ogilvie, P., Callan, J.P.: combining document representations for known-item search. In: SIGIR 2003, pp. 143–150 (2003)

16. Page, L., Brin, S., Motwani, R., Winograd, T.: The PageRank citation ranking: bringing order to the web. Technical report 1999-66, Stanford InfoLab, November 1999

17. Pound, J., Mika, P., Zaragoza, H.: Ad-hoc object retrieval in the web of data. In: WWW 2010, pp. 771–780 (2010)

18. Robertson, S.E., Zaragoza, H.: The probabilistic relevance framework: BM25 and beyond. Found. Trends Inf. Retr. 3(4), 333–389 (2009)

19. Sinha, V., Karger, D.R.: Magnet: supporting navigation in semistructured data environments. In: SIGMOD 2005, pp. 97–106 (2005)

20. Varadarajan, R., Hristidis, V., Raschid, L.: Explaining and reformulating authority flow queries. In: ICDE 2008, pp. 883–892 (2008)

21. Zhai, C.: Statistical language models for information retrieval: a critical review. Found. Trends Inf. Retr. 2(3), 137–213 (2008)

A Ranking Based Approach for Robust Object Discovery from Images of Mixed Classes

Min Ge$^{(\boxtimes)}$, Chenyi Zhuang, and Qiang Ma

Kyoto University, Kyoto, Japan
gemin@db.soc.i.kyoto-u.ac.jp

Abstract. Discovering knowledge from social images available on social network services (SNSs) is in the spotlight. For example, objects that appear frequently in images shot around a certain city may represent its characteristics (local culture, etc.) and may become the valuable sightseeing resources for people from other countries or cities. However, due to the diverse quality of social images, it is still not easy to discover such common objects from them with the conventional object discovery methods. In this paper, we propose a novel unsupervised ranking method of predicted object bounding boxes for discovering common objects from a mixed-class and noisy image dataset. Extensive experiments on standard and extended benchmarks demonstrate the effectiveness of our proposed approach. We also show the usefulness of our method with a real application in which a city's characteristics (i.e., culture elements) are discovered from a set of images collected there.

1 Introduction

With the rapid development of SNSs, a large number of social images are generated every day. Various kinds of knowledge can be mined from social images. For example, by mining social images, methods of revealing the characteristics of visitors and spots have been well studied [1–3] and such kind of knowledge can be used for discovering sightseeing resources from the user viewpoints rather than that of the providers (travel agencies, etc.).

One of the key points of social images mining is how to handle its diverse quality, i.e., social images are always mixed with different kinds of objects and noisy images. Doersch et al. point out that the common objects, such as patches from street-view images are representative objects of a city characteristics [4]. Meanwhile, the interpretability of their image patches needs to be improved. Therefore, there is a challenge of discovering interpretative common objects from mixed-class and noisy images.[1] Although there are some existing unsupervised object discovery approaches that can be used to discover common objects even in noisy and mixed-class image collections [5–8], most of them are unable to distinguish between common objects and noisy images. In addition, due to the

---

[1] Without loss of generality, hereafter, we generalize social image dataset as a set of mixed-class and noisy images.

© Springer International Publishing AG 2017
W.-K. Sung et al. (Eds.): AIRS 2017, LNCS 10648, pp. 71–83, 2017.
https://doi.org/10.1007/978-3-319-70145-5_6

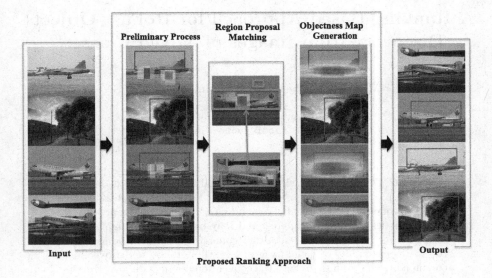

Fig. 1. Process flow of our proposed approach.

number of noisy images that are unknown in advance, a robust approach with high accuracy is necessary.

In this paper, we propose a novel unsupervised ranking approach for improving the performance of object discovery from images of mixed class. Our proposed approach can be applied to any existing unsupervised object discovery methods to improve both the robustness and accuracy.

The major contributions of this paper can be summarized as follows.

1. A novel region proposal matching approach and objectness map based ranking approach are proposed. Our ranking approach is sufficiently robust to remove object bounding boxes that appear on noisy images (Sect. 3).
2. Extensive experiments are conducted on two challenging benchmarks. The experimental results show that our approach improves the performance of related unsupervised object discovery significantly (Sect. 4).
3. The high ranked bounding boxes can be used for various applications. To verify the effectiveness of our proposed approach, we apply our approach to a real world application. In this application scenario, we discovered the objects or parts that mostly represent a city's characteristics which are quite useful for discovering various sightseeing resources (Sect. 5).

2   Related Work

Supervised object detection is a challenging and important task which can be used in many applications. With the rapid development of deep learning, the performance of object detection approaches has been improved dramatically [9–11]. However, these approaches detect objects in a supervised manner. It means that a large amount of manually labeled training data is necessary to train the neural network, which is time-consuming.

Unsupervised object discovery is more challenging because of the lack of annotation information for training. Tang et al. [7] presented a joint image–box formulation for object discovery. Cho et al. [5] tackled the discovery and localization problem using probabilistic Hough matching. There are also some approaches that attempt to discover objects in videos [12]. In this study, we rank the predicted bounding boxes discovered by existing approaches to improve the robustness and accuracy.

Objectness measurement [13,14] is often used for the initialization of object discovery. We also propose a objectness map, which is used for ranking object bounding boxes. However, our approach takes the discriminative patch and region proposal similarity into consideration and is more robust for removing noisy images.

3 Proposed Approach

3.1 Overview

As shown in Fig. 1, the input of our approach is a set of images and output is a list of ranked predicted bounding boxes. At first, we apply a certain unsupervised object discovery method to obtain the predicted bounding boxes. Meanwhile, we detect the discriminative patches of each image, which are cues to the location of potential common objects. After that, to refine the candidates of common objects, we extract and match region proposals in images based on their appearance similarity and detected discriminative patches. Matching scores will be assigned to region proposals and their pixels. As a result, for each image, we can generate an objectness map to indicate the probability (objectness score) that a pixel belongs to an object. Finally, we assign each predicated object bounding box a ranking score based on the intersection with the objectness map. As the result, we can select a subset of bounding boxes with high object discovery performance and little noise.

3.2 Preliminary Process

Discriminative patch is a type of mid-level visual element that depicts a small part of an object. We discover discriminative patches by applying the approach proposed by Singh et al. [15]. For a given image I, P is the set of discovered patches. $p_i \in P$ is the ith discovered patch and its SVM confidence score is denoted by S_i. For further processing, without loss of generality, each S_i is translated into a positive value. Figure 2 shows four examples of the heat maps generated by 10 SVM detectors with highest confidence score in each image.

Region proposals are widely used both in supervised and unsupervised object localization approaches which restrict the search space of possible object bounding boxes. We apply a well-known method [16] to extract region proposals. The set of discovered region proposals for image I is denoted by R. $r \in R$ is one region proposal coming from set R.

Fig. 2. Four examples of the heat map generated by top 10 SVM detectors.

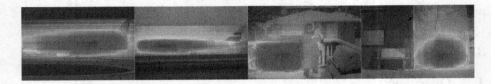

Fig. 3. Four examples of the objectness map after accumulation.

3.3   Region Proposal Matching

Region proposal matching is quite effective for unsupervised object discovery [5] but it can not distinguish between common objects and noisy images. Discriminative patch is a kind of robust mid-level visual element representation. Benefit from the training of SVM detector, discriminative patch is able to discover common object parts from a very noisy image set. However, because of the limitation of size and shape, discriminative patch is not suitable for unsupervised object discover and bounding box ranking. As shown in Fig. 2, discriminative patches only covers a very small part of common objects and some of them even appears on the background.

In this section, we propose a novel matching process by combining the advantages of region proposal matching and discriminative patch. We first define the discriminativity of each region proposals found in images which shows the probability that the region proposal is depicting a domain object rather than a noise one. Then we retrieval similar image pairs and define matching probability for regions in each image pair. The matching probability both takes discriminativity and appearance similarity of region proposal into consideration to improve the accuracy performance. Finally, we give a final matching score to each region proposals of each image.

Discriminativity of Region. For an input image I, L_i is a set of pixels that are contained in patch p_i. We only consider discriminative patches that have the top k ($k = 10$ in our current work) SVM confidence scores. These discriminative patches will be used to draw a heat map M. The value of each pixel l in M can be calculated as follows.

$$M(l) = \sum S_i \quad \text{where } l \in L_i \text{ for } i = 1, \ldots, 10. \tag{1}$$

Note that a pixel in M will be given a zero value if it is not contained in any discovered discriminative patches.

We use $p(D|r)$ to denote discriminativity probability that region proposal r is depicting an object D determined by discriminative patches. The idea behind $p(D|r)$ is that one region proposal should have a high probability of depicting an object if it contains a significant number of pixels with high values and less pixels with zero values in the heat map. To obtain tight region proposals, we introduce a penalty value to each pixel l that $M(l) = 0$. We modify the definition of $M(l)$ as follows.

$$M'(l) = \begin{cases} M(l) & \text{if } M(l) > 0 \\ -\frac{\sum_{l \in I} M(l)}{|I|} & \text{if } M(l) = 0 \end{cases} \tag{2}$$

$|I|$ is the number of pixels in image I. Then we calculate $p(D|r)$

$$p(D|r) = sign\left(\sum_{l \in r} M'(l) \right). \tag{3}$$

The function *sign* transforms the value range to $(0, 1)$.

Matching Probability. In this part, We take both the appearance similarity and discriminativity into consideration. The appearance similarity denotes the similarity on the basis of visual features extracted from the two target regions. We first calculate the matching probability of two regions r_1 and r_2 in two images. The probability of matching is denoted by $p(r_1 = r_2|r_1, r_2)$. We assume that if r_1 and r_2 match well, they should also contain the foreground object determined by discriminative patches, which implies that $p(r_1 = r_2, D|r_1, r_2)$. Based on this assumption, the matching probability can be written as follows.

$$\begin{aligned} p(r_1 = r_2|r_1, r_2) &= p(r_1 = r_2, D|r_1, r_2) \\ &= p(r_1 = r_2|D, r_1, r_2)p(D|r_1, r_2) \\ &= p(r_1 = r_2|D, r_1, r_2)p(D|r_1)p(D|r_2). \end{aligned} \tag{4}$$

Here, $p(r_1 = r_2|D, r_1, r_2)$ can be denoted by the similarity of appearance of region r_1 and r_2 which is calculated the cosine distance of HOG feature [19].

Matching Score. A higher matching score indicates a higher possibility of the region proposal depicting an object. In practice, we retrieve the 10 nearest neighbors for each single target image. The matching score of the region r_i in the given image can be calculated as follows.

$$c(r_i) = \sum_k \max(p(r_i = r_{kj}|r_i, r_{kj})), \tag{5}$$

where r_{kj} is the jth region in the kth nearest neighbor of the target image. We assign a matching score to each region proposal of each image.

3.4   Objectness Map Generation

The region proposals with matching scores cannot be directly used for ranking object bounding boxes for two reasons.

1. The accuracy is not sufficiently high to improve the performance effectively. The global and local features cannot describe the content in images and region proposals perfectly, which will lead to incorrect nearest neighbors and errors in matching. Some region proposals may be quite similar visually, but they may be quite different from each other in terms of the feature space.
2. Region proposals cannot be used for ranking. Our goal is to improve the performance of previous object discovery methods by ranking the bounding boxes. However, region proposals are only a very small subset of all possible bounding boxes and sometimes discovered bounding boxes do not belong to any region proposal used in Sect. 3.2.

We further introduce a pixel level objectness map for ranking objects. The pixel-level objectness map is the regularized accumulation of matching scores of region proposals. The idea is that if a single pixel belongs to a foreground object, it is highly possible that the region proposals containing this pixel should have high matching scores. Although some of the matching scores are incorrect, it is very rare that all the matchings are wrong. In short, region proposals will vote for every pixel contained in an image. We collect all votes coming from all region proposals and generate a pixel-level objectness map for each image.

Suppose that A is the objectness map for one image and l is a single pixel in this image. Then the value of each pixel l in A can be calculated as follows.

$$A(l) = \frac{\sum_{r_i \in R_l} c(r_i)}{\sum_{r_j \in R} c(r_j)}, \tag{6}$$

where R_l is the set of regions containing the pixel l. Figure 3 shows some examples of objectness maps.

3.5   Ranking

We use the objectness map for ranking. If the pixels in a predicted object bounding box have very high values in the objectness map, it is highly probable that

Fig. 4. The images in the first column are original images. The second column shows the objectness maps generated by our approach. The third column shows the bounding boxes discovered by [5]. The bounding box in the first row covers very few pixels that have high objectness score. Therefore, the ranking score given to this bounding box will be quite low. In contrast, the ranking score of the bounding box in the second row will be quite high since it covers most of the high objectness pixels.

the object bounding box is the object that we want to discover. Figure 4 shows some examples that demonstrate this idea.

Given an image I, A is the corresponding objectness map and r is the discovered object bounding box. The ranking score s is calculated by the sum of the map value of pixels that are contained in r, in other words, $s_r = \sum A(l)$ for all $l \in r$. Here, l is a pixel in image I. s_r could be treated as the score of objectness that can be used for ranking region proposals. However, s_r is not robust enough to remove noisy images. In Sect. 3.2, we introduced the discriminative probability $p(D|r)$ for matching region proposals, which is capable of distinguishing domain objects and noisy images. Therefore, we take both s_r and $p(D|r)$ into consideration for the final ranking score. The final ranking score DS of region proposal r is defined as follows.

$$DS(r) = s_r \times p(D|r). \tag{7}$$

4 Experimental Evaluation

We verified our model on two benchmark datasets and their extended datasets. All the experiments were conducted in an unsupervised manner. Several baseline methods are implemented for comparison.

4.1 Evaluation Metrics

The correct localization (CorLoc) metric is widely used in object discovery [5,7]. It is defined as the percentage of images correctly localized according to the Pascal criterion. A predicted object bounding box is regarded as correct if the ratio of the intersection area and union area of the predicted bounding box and ground truth is larger than 0.5. Since our goal is to rank the discovered object bounding boxes to help output a subset of bounding boxes with best CorLoc performance, we only concern with the CorLoc performance at top k bounding boxes (k varies from 10 to 400).

In addition, the resistance to noise of our proposed approach is also verified. Purity is calculated for each extended dataset. For a particular k, purity is calculated as the ratio of the number of images that belong to the standard benchmark and the value of k. A high purity score means high resistance to noise.

4.2   Implementation Detail and Baselines

The training of discriminative patch needs positive and negative datasets. The positive set for training is all the images in the dataset and the negative set contains the random images downloaded from Flickr. There is no manual pre-processing on the negative set. GIST [18] is used as the global feature for the retrieval of the nearest neighbors and HOG [19] is used as the local feature for each region proposal (Sect. 3.2). The initial predicted object bounding boxes are discovered using the approach proposed in [5].

We compared our method with three baselines: (1) UODL: Predicted object bounding boxes are ranked by the confidence score given by the method of [5]. (2) SM: We replace our objectness map with a saliency map [14] in the ranking step. (3) OBJ: We rank the bounding boxes using the objectness score given by the approach from [13].

4.3   PASCAL VOC 2007 Dataset

DataSet. Our experiment was conducted on a pure and standard benchmark and its extended datasets with different ratio of noisy images

1. Standard benchmark data set: Pascal VOC 2007 [17] is a well-known and challenging image dataset that contains 20 different object classes. Our experiments are conducted on a subset of Pascal VOC 2007 called Pascal07-6x2, which contains 6 object classes, 12 viewpoints and 463 images. There is at least one object in each single image. Therefore, there are no noisy images in this dataset.
2. Extended data sets with noise: To evaluate the performance of our approach on a noisy dataset, we extend the original Pascal07-6x2 dataset and add noisy images. Noisy images are random images downloaded from Flickr[2]. After manual preprocessing, we guarantee that there are no common objects in the noisy images. There are five extended datasets in our experiments with different noise ratios from 0.1 to 0.5. (the ratio of noisy images and images from Pascal07-6x2).

---

[2] https://www.flickr.com/.

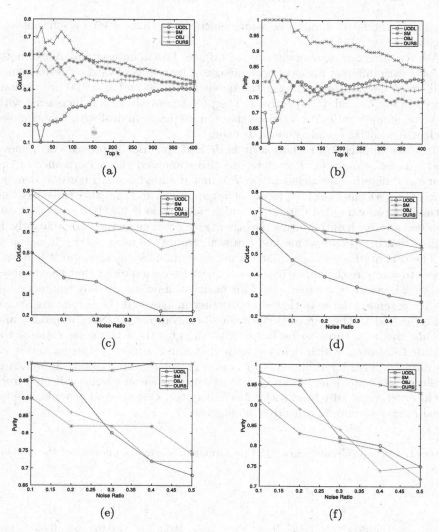

Fig. 5. Performance comparison of UODL, SM, OBJ and our proposed approach: (a) CorLoc performance on a dataset with noise ratio 0.4; (b) Purity performance on a dataset with noise ratio 0.4; (c) Top 50 CorLoc performance; (d) Top 100 CorLoc performance; (e) Top 50 purity performance; (f) Top 100 purity performance.

Experimental Results. Figure 5a shows the CorLoc performance of our proposed approach and three baseline methods on an extended dataset of which noise ratio is 0.4. Our approach outperforms the other three baseline methods significantly when k is smaller than 400. When k is very small, for example when $k = 20$, our approach achieves the best performance (CorLoc = 75%). When k is larger than 400, the performance of our proposed approach and the other two baseline methods are almost the same, which shows that ranking is efficient and necessary.

We repeated our experiments on the original Pascal dataset and all extended datasets with different ratios of noisy images. The results of CorLoc performance of different methods on all six datasets when $k = 50$ and $k = 100$ are demonstrated in Fig. 5c and d, respectively. The performance margin is increasing with the value of noisy ratio. It is obvious that our method can deal with noisy images efficiently, while the related methods cannot.

In some application scenarios, it is desired that our approach can remove noisy images in outputs. Therefore, we also compared the performance of the purity of different approaches in our conducted experiments. Figure 5b demonstrates the performance of the purity of our approach and the other three baseline methods on an extended dataset of which noise ratio is 0.4. It reveals that our proposed approach outperforms the other three baseline methods significantly. It verified that our proposed method can well remove the noisy images in outputs.

The original Pascal07-6x2 dataset contains no noisy images and it is meaningless to compare the purity of different methods on a dataset that has no noisy images. Therefore, the experiments for purity evaluation are only conducted on the five extended datasets that contain noisy images and the results are shown in Fig. 5e and f. The gap of performance between our proposed approach and baseline methods turns to be larger with the growth of the noise ratio of the extended datasets, which demonstrates that our approach performs better on the datasets with large amounts of noise. Therefore, our ranking approach is applicable to many kinds of noisy datasets such as social images collected from the Internet, especially from social media services. Our method is applicable for social images to mining collective intelligences.

Table 1. CorLoc (C) and Purity (P) performance of each component of the ranking score

Noise ratio	0		0.1		0.2		0.3		0.4		0.5	
	C	P	C	P	C	P	C	P	C	P	C	P
s_r	**0.86**	NaN	0.68	0.88	0.58	0.86	0.62	0.8	0.62	0.8	0.56	0.68
$p(D|r)$	0.66	NaN	0.58	0.94	0.46	**0.98**	0.52	0.94	0.48	0.9	0.54	0.94
$DS(r)$	0.64	NaN	**0.76**	1	**0.68**	0.98	**0.66**	0.98	**0.66**	1	**0.64**	1

We also analyze each component of the final ranking score which is shown in Table 1. The accuracy of s_r is better than that of $p(D|r)$. $p(D|r)$ is more robust than s_r (better purity performance). $DS(r)$ achieves the best performance of accuracy and robustness.

4.4 Object Discovery Dataset

The object discovery dataset [6] was collected by Bing API. We use a subset of 300 images which consist of 3 classes, airplane, car and horse. In order to verify the robustness of our approach, we add 150 random images to the dataset.

Table 2. CorLoc (C) and Purity (P) performance on Object Discovery Dataset with additional 150 noise images

Top k	50		100		150		200	
	C	P	C	P	C	P	C	P
UODL	0.62	0.84	0.64	0.84	0.613	0.8	0.595	0.795
SM	0.4	0.42	0.47	0.51	0.493	0.56	0.51	0.6
OBJ	0.62	0.68	0.64	0.7	0.573	0.687	0.565	0.69
Ours	**0.68**	**0.9**	**0.73**	**0.91**	**0.72**	**0.88**	**0.66**	**0.84**

Table 2 shows the performance of our approach and baselines when we retrieved top k images and k varies from 50 to 200. The accuracy and robustness performance of our approach outperforms all the other baseline methods.

5 Application

To verify the utility of our method, we further applied it in a real world application. In this application, our ranking method was used for the visual characteristic discovery of sightseeing spots in a city. Our dataset consists of photos of sightseeing spots from five cities: Kyoto, Xi'an, Beijing, Paris, and San Francisco. We first selected about 30 most popular sightseeing spots from TripAdvisor[3] for each city and downloaded 50–100 images for each sightseeing spot and sampled 1000 images randomly for each city. We regarded the 1000 images taken in Kyoto as a positive set, whereas the images from other cities and a small image set of random nature scenes as a negative set to discover discriminative patches. No manual processing is performed on the dataset.

After obtaining the positive and negative image sets, we applied our method to discover discriminative visual elements (culture elements) in Kyoto. Figure 6 shows the top 20 discovered visual elements. Note that, although the positive dataset contains a large number of noisy images (more than 50%), our proposed method is still able to output a relatively pure subset. In particular, in the top 10 images, almost all the images in this subset depict some of the characteristics of sightseeing spots in Kyoto such as temples, towers, traditional houses, and maple trees. Characteristics mined from sightseeing spot are quite useful for discovering new sightseeing spots, especially the sightseeing spots that are not very well-known. Such kind of sightseeing spots will be quite attractive for the travellers that desire for personalized and deep travel.

---

[3] https://www.tripadvisor.com/.

Fig. 6. Top 20 predicted bounding boxes discovered from the image dataset of Kyoto

6 Conclusion

In this paper, we proposed a novel approach for ranking object bounding boxes in an unsupervised manner from mixed-class and noisy images. Extensive experiments and an evaluation of the applicability have demonstrated the effectiveness and robustness of our approach to rank the object bounding boxes. The experimental results and example application reveal that our approach is helpful to discover meaningful knowledge from social images.

Acknowledgement. This work is partly supported by JSPS KAKENHI (16K12532) and MIC SCOPE (172307001).

References

1. Zhuang, C., Ma, Q., Liang, X., Yoshikawa, M.: Discovering obscure sightseeing spots by analysis of geo-tagged social images. In: ASONAM (2015)
2. Zhuang, C., Ma, Q., Yoshikawa, M.: SNS user classification and its application to obscure POI discovery. MTA **76**(4), 5461–5487 (2016)
3. Shen, Y., Ge, M., Zhuang, C., Ma, Q.: Sightseeing value estimation by analyzing geosocial images. In: BigMM (2016)
4. Doersch, C., Singh, S., Gupta, A., Sivic, J., Efros, A.: What makes Paris look like Paris? TOG **31**(4) (2012)
5. Cho, M., Kwak, S., Schmid, C., Ponce, J.: Unsupervised object discovery and localization in the wild: part-based matching with bottom-up region proposals. In: CVPR (2015)
6. Rubinstein, M., Joulin, A., Kopf, J., Liu, C.: Unsupervised joint object discovery and segmentation in internet images. In: CVPR (2013)
7. Tang, K., Joulin, A., Li, L.J., Fei-Fei, L.: Co-localization in real-world images. In: CVPR (2014)
8. Vahdat, A., Mori, G.: Handling uncertain tags in visual recognition. In: ICCV (2013)
9. Girshick, R., Donahue, J., Darrell, T., Malik, J.: Rich feature hierarchies for accurate object detection and semantic segmentation. In: CVPR (2014)
10. Girshick, R.: Fast R-CNN. In: ICCV (2015)
11. He, K., Zhang, X., Ren, S., Sun, J.: Spatial pyramid pooling in deep convolutional networks for visual recognition. In: Fleet, D., Pajdla, T., Schiele, B., Tuytelaars, T. (eds.) ECCV 2014. LNCS, vol. 8691, pp. 346–361. Springer, Cham (2014). doi:10.1007/978-3-319-10578-9_23
12. Kwak, S., Cho, M., Laptev, I., Ponce, J., Schmid, C.: Unsupervised object discovery and tracking in video collections. In: ICCV (2015)

13. Alexe, B., Deselaers, T., Ferrari, V.: Measuring the objectness of image windows. TPAMI **34**(11), 2189–2202 (2012)
14. Harel, J., Koch, C., Perona, P.: Graph-based visual saliency. NIPS **1**(2), 5 (2006)
15. Singh, S., Gupta, A., Efros, A.A.: Unsupervised discovery of mid-level discriminative patches. In: Fitzgibbon, A., Lazebnik, S., Perona, P., Sato, Y., Schmid, C. (eds.) ECCV 2012. LNCS, pp. 73–86. Springer, Heidelberg (2012). doi:10.1007/978-3-642-33709-3_6
16. Manen, S., Guillaumin, M., Van Gool, L.: Prime object proposals with randomized prim's algorithm. In: ICCV (2013)
17. Everingham, M., Van Gool, L., Williams, C.K., Winn, J., Zisserman, A.: The PASCAL visual object classes challenge 2007 (VOC 2007) results. In: Citeseer (2007)
18. Torralba, A., Fergus, R., Weiss, Y.: Small codes and large image databases for recognition. In: CVPR (2008)
19. Dalal, N., Triggs, B.: Histograms of oriented gradients for human detection. In: CVPR (2005)

Fast Exact Algorithm to Solve Continuous Similarity Search for Evolving Queries

Tomohiro Yamazaki[✉], Hisashi Koga, and Takahisa Toda

Graduate School of Informatics and Engineering,
The University of Electro-Communications,
Chofugaoka 1-5-1, Chofu, Tokyo 182-8585, Japan
yamazaki@sd.is.uec.ac.jp, {koga,takahisa.toda}@is.uec.ac.jp

Abstract. We study the continuous similarity search problem for evolving queries which has recently been formulated. Given a data stream and a database composed of n sets of items, the purpose of this problem is to maintain the top-k most similar sets to the query which evolves over time and consists of the latest W items in the data stream. For this problem, the previous exact algorithm adopts a pruning strategy which, at the present time T, decides the candidates of the top-k most similar sets from past similarity values and computes the similarity values only for them. This paper proposes a new exact algorithm which shortens the execution time by computing the similarity values only for sets whose similarity values at T can change from time $T-1$. We identify such sets very fast with frequency-based inverted lists (FIL). Moreover, we derive the similarity values at T in $O(1)$ time by updating the previous values computed at time $T-1$. Experimentally, our exact algorithm runs faster than the previous exact algorithm by one order of magnitude.

Keywords: Data stream · Evolving query · Set similarity search · Inverted lists

1 Introduction

With the prosperity of Internet of Things (IoT), analysis of data streams becomes more significant. Especially, the similarity search problem for data streams has attracted much attention, because it is useful for many applications such as information recommendation and anomaly detection.

While many similarity search problems have been defined for data streams, this paper studies the Continuous Similarity Search Problem for Evolving Queries (abbreviated as CSPEQ hereafter) recently formulated by Xu *et al.* [8]. The CSPEQ deals with a data stream to which one alphabet is added at every time instant. While the latest W items in the data stream become the query, the task of CSPEQ is to continuously seek the top-k most similar sets to the query from the static database composed of n sets of alphabets. Thus, every time a new item is added to the stream, the query evolves and we must update the top-k most similar sets to the evolving query. The CSPEQ models a scenario

© Springer International Publishing AG 2017
W.-K. Sung et al. (Eds.): AIRS 2017, LNCS 10648, pp. 84–96, 2017.
https://doi.org/10.1007/978-3-319-70145-5_7

in which the recommendation system must adapt to the changing preference of a user. The evolving query corresponds to the changing preference of the user, whereas the top-k most similar sets abstract the objects recommended to the user.

The brute-force algorithm for the CSPEQ is to compute the similarity values between the query and all the sets in the database at every time instant. To reduce the frequency of similarity computation, Xu *et al.* [8] proposed a pruning-based exact algorithm. This algorithm judges if a set S is a candidate of the top-k most similar sets to the query at time T from some past similarity value for S. Then, the similarity value is actually computed for S at T, if S is a candidate. They also proposed another approximation algorithm based on Minhash [1].

In this paper, taking a different approach from [8], we develop a new exact algorithm. Our algorithm shortens the execution time by computing the similarity values only for sets whose similarity values at T can change from time $T - 1$. We identify such sets very fast through novel frequency-based inverted lists (FIL). To make the execution time shorter, our algorithm obtains the similarity values at T by updating the corresponding similarity values at $T - 1$ in $O(1)$ time. Our main contributions are twofold and stated as follows:

(1) We show that the CSPEQ is solved efficiently with the inverted lists, despite the inverted lists appear difficult to be applied to the CSPEQ, as will be explained later.
(2) Our exact algorithm experimentally runs faster than the previous exact algorithm in [8] by one order of magnitude. Indeed, it runs as fast as the previous approximation algorithm.

This paper is organized as follows. Section 2 formally defines the CSPEQ problem. Section 3 describes the previous methods for the CSPEQ. Then, Sect. 4 presents our method. Section 5 reports the experimental evaluation. Section 6 reviews related works. Section 7 concludes this paper.

2   Problem Statement

This section formally defines the CSPEQ problem defined in [8]. Let $\Phi = \{x_1, x_2, \cdots, x_{|\Phi|}\}$ be a set of alphabets. To the data stream, exactly one new element from Φ is added at every time instant. We denote an element added at time T by e_T. In the CSPEQ, the latest W items in the data stream become the query Q_T at time T. That is, $Q_T = \{e_{T-W+1}, e_{T-W+2}, \cdots, e_T\}$. On the other hand, the database D manages n sets of alphabets $\{S_1, S_2, \cdots, S_n\}$ all of which are static and do not change over time. The task of CSPEQ is to search for the most similar k (top-k) sets to Q_T in D at every T. Here, we evaluate the similarity between the query Q_T and a set S in D with the Jaccard similarity in Eq. (1).

$$\text{sim}(S, Q_T) = \frac{|S \cap Q_T|}{|S \cup Q_T|}. \tag{1}$$

The CSPEQ allows S and Q_T to be multisets. When S or Q_T is a multiset, the CSPEQ uses the extended Jaccard similarity in which $|S \cap Q_T| = \sum_{i=1}^{|\Phi|} \min(s_i, q_i)$ and $|S \cup Q_T| = \sum_{i=1}^{|\Phi|} \max(s_i, q_i)$. Here, s_i and q_i symbolize the numbers of the i-th alphabet x_i in S and Q_T, respectively.

The brute-force algorithm for the CSPEQ mentioned in Sect. 1 often becomes too slow for practical use, because it has to calculate $\mathrm{sim}(S, Q_T)$ for every set S in D at every time instant and it takes an $O(|Q_T| + |S|) = O(W + |S|)$ time to compute $\mathrm{sim}(S, Q_T)$.

3 Previous Methods

This section reviews the previous exact algorithm based on a pruning technique in Sect. 3.1 and the previous approximation algorithm based on Minhash in Sect. 3.2. They were both developed in [8].

3.1 Pruning-Based Exact Algorithm

With a pruning technique, the exact algorithm named GP (General Pruning-based) in [8] saves the similarity computation between the query and the sets in D. At time T, GP first chooses a group of sets $R \subset D$ which contain more than k sets and determines the top-k similar sets $\{R_1, R_2, \cdots R_k\}$ in R, where R_k presents the k-th most similar set in R. Let $\tau = \mathrm{sim}(R_k, Q_T)$. As $R \subset D$, τ gives a lower bound for the similarity value between Q_T and the k-th most similar set in D. As for a set $S \in D \backslash R$, GP calculates an upper bound of $\mathrm{sim}(S, Q_T)$ from the similarity value computed most recently before T for S. Then, GP avoids computing $\mathrm{sim}(S, Q_T)$, if this upper bound is less than τ. Thus, GP omits the similarity computation for some sets in $D \backslash R$. Since R influences the pruning performance, GP chooses R in a sophisticated way. Refer to [8] for details.

The upper bound of $\mathrm{sim}(S, Q_T)$ is derived as $\frac{(|S|+|Q_t|+T-t)\cdot\mathrm{sim}(S,Q_t)+(T-t)}{|S|+|Q_t|-(T-t)\cdot\mathrm{sim}(S,Q_t)-(T-t)}$ in [8], where t denotes the time at which the similarity value of S was computed for the last time before T. Thus, this upper bound can be easily computed from the known $\mathrm{sim}(S, Q_t)$.

The primary characteristics of GP are that (1) GP decides whether to compute $\mathrm{sim}(S, Q_T)$ based on the magnitude of the past similarity value for S and that (2) GP has to compute the Jaccard similarity $\mathrm{sim}(S, Q_T)$ in $O(W + |S|)$ time, when GP needs to compute it.

3.2 Approximation Algorithm

Xu *et al.* [8] also presented an approximation algorithm named MHI (Minhash-based algorithm using inverted indices). MHI runs faster than GP.

Minhash is a randomized hash function useful for set similarity search with the Jaccard similarity. For Minhash, the collision probability between two sets A and B equals their Jaccard similarity. Thus, $\mathrm{P}[h(A) = h(B)] = \frac{|A \cap B|}{|A \cup B|} =$

$sim(A, B)$. Therefore, if we prepare r distinct hash functions $\{h_1, h_2, \cdots, h_r\}$, $sim(S, Q_T)$ can be approximated by $\frac{\text{\# of hash functions s.t. } h(S) = h(Q_T)}{r}$. For brevity, we refer to the term "approximate similarity" as "asim" here.

MHI obtains the asim values at time T by updating the old asim values at time $T - 1$ as follows. At T, MHI first computes the r hash values from $h_1(Q_T)$ to $h_r(Q_T)$ for the new query Q_T. Then, for $1 \leq i \leq r$, the asim of S is increased by $\frac{1}{r}$ at T, if $h_i(S) = h_i(Q_T)$. On the contrary, if $h_i(S)$ was equal to $h_i(Q_{T-1})$ at $T - 1$, the asim of S is decreased by $\frac{1}{r}$ at T. Note that, to obtain the asim values for all the sets in D at T, MHI has only to access the sets S which satisfy either $h_i(S) = h_i(Q_{T-1})$ or $h_i(S) = h_i(Q_T)$ for $1 \leq i \leq r$. These sets are discovered efficiently via hash tables, i.e., inverted lists regarding hash values.

MHI is similar to our algorithm, because it relies on the inverted lists and the new asim values at time T are derived by updating the old asim values at time $T - 1$. However, our method differs from MHI in the following way.

(1) Though MHI is an approximation algorithm, our method is an exact algorithm. To improve the accuracy, MHI must increase the number of hash functions r, with sacrificing both the time complexity and the space complexity. Our method is free from such a trade-off. Moreover, MHI cannot treat multisets properly, since Minhash does not support them. Our method handles multisets without problem.
(2) Applying the inverted lists to MHI is more straightforward as compared to our case. For example, MHI need not care about the frequency of items in the inverted lists.

4 Proposed Method

At T, GP decides not to compute $sim(S, Q_T)$ according to the magnitude of the past similarity value of S. Differently, our method will skip computing $sim(S, Q_T)$, in case $sim(S, Q_T)$ does not change from the previous $sim(S, Q_{T-1})$. Remarkably, when our method has to compute $sim(S, Q_T)$, $sim(S, Q_T)$ is derived instantly by updating $sim(S, Q_{T-1})$ in $O(1)$ time.

4.1 Relationship Between $sim(S, Q_{T-1})$ and $sim(S, Q_T)$

Before describing our algorithm, we consider the relation between $sim(S, Q_{T-1})$ and $sim(S, Q_T)$ for a set S. Particularly, we address the relation about intersection size and the relation about union size separately. Note that they serve as the numerator and the denominator of the Jaccard similarity, respectively.

At T, we have $Q_T = \{e_{T-W+1}, e_{T-W+2}, \cdots, e_{T-1}, e_T\}$. Q_T is built by adding e_T to Q_{T-1} and ejecting e_{T-W} from Q_{T-1}. Denote e_T as IN and e_{T-W} as OUT. Let Q'_T be the substring shared by Q_{T-1} and Q_T, i.e., $Q'_T = Q_{T-1} \backslash OUT = Q_T \backslash IN$.

When Q_{T-1} evolves to Q_T, we may assume, without losing the accuracy of discussion about the intersection size and the union size, that Q_{T-1} first becomes Q'_T by discarding OUT, and then, Q'_T changes to Q_T after IN joins. This conversion process is depicted as $Q_{T-1} \to Q'_T \to Q_T$.

Regarding $Q'_T \to Q_T$, Lemmas 1 and 2 hold. Lemma 1 treats the intersection size, while Lemma 2 discusses the union size.

Lemma 1. *If $IN \in S \backslash Q'_T$, $|S \cap Q_T| = |S \cap Q'_T| + 1$. Otherwise, $|S \cap Q_T| = |S \cap Q'_T|$.*

Proof. Note that $S \backslash Q'_T$ contains the elements in S which do not belong to the intersection $S \cap Q'_T$. Therefore, if IN arrives at T and $IN \in S \backslash Q'_T$, the intersection size increases by just 1. □

Lemma 2. *If $IN \notin S \backslash Q'_T$, $|S \cup Q_T| = |S \cup Q'_T| + 1$. Otherwise, $|S \cup Q_T| = |S \cup Q'_T|$.*

Proof. Since $|S \cup Q_T| = |S| + |Q_T| - |S \cap Q_T|$ and $|Q_T| = |Q'_T| + 1$, it holds that $|S \cup Q_T| = |S| + |Q'_T| + 1 - |S \cap Q_T|$. Lemma 2 is derived by rewriting the term $|S \cap Q_T|$ in this equation according to Lemma 1. □

Regarding $Q'_T \to Q_T$, Lemmas 3 and 4 hold.

Lemma 3. *If $OUT \in S \backslash Q'_T$, $|S \cap Q'_T| = |S \cap Q_{T-1}| - 1$. Otherwise, $|S \cap Q'_T| = |S \cap Q_{T-1}|$.*

Lemma 4. *If $OUT \notin S \backslash Q'_T$, $|S \cup Q'_T| = |S \cup Q_{T-1}| - 1$. Otherwise, $|S \cup Q'_T| = |S \cup Q_{T-1}|$.*

Proof of Lemmas 3 and 4. Q_{T-1} is the same as Q_T in that it is generated by adding one element to Q'_T. Therefore, Lemmas 1 and 2 still hold, if we replace Q_T with Q_{T-1} and IN with OUT. Thus, if $OUT \in S \backslash Q'_T$, $|S \cap Q_{T-1}| = |S \cap Q'_T| + 1$ and if $OUT \notin S \backslash Q'_T$, $|S \cup Q_{T-1}| = |S \cup Q'_T| + 1$. □

If we regard these 4 lemmas as the construction rules, we can design an algorithm which measures the similarity values between Q_T and all the sets in D by updating their similarity values to Q_{T-1}. Figure 1 illustrates a simple algorithm which embodies this policy. We call this algorithm as SIMPLE. SIMPLE always memorizes the intersection size and the union size between the query and any set S in D in the variables S.intersection and S.union. SIMPLE assumes that S.intersection $= |S \cap Q_{T-1}|$ and S.union $= |S \cup Q_{T-1}|$ before the execution. After the execution, it holds that S.intersection $= |S \cap Q_T|$ and S.union $= |S \cup Q_T|$. After SIMPLE finishes, the top-k most similar sets to Q_T can be obviously selected, since we have $\text{sim}(S, Q_T)$ for every $S \in D$. This selection takes an $O(n \log k)$ time, if the minimum heap is utilized.

Unfortunately, SIMPLE is slow, since it processes all the sets in D and it must check whether IN and OUT belong to $S \backslash Q'_T$. Normally, this check takes an $O(|S| + |Q'_T|) = O(|S| + W)$ time.

Our algorithm in Sect. 4.3 enhances SIMPLE to improve the execution speed with inverted lists. However, the application of inverted lists to SIMPLE is not trivial for the following reason: The standard framework of inverted lists arranges

1: **foreach** $S \in D$ **do**
2:     **if** $OUT \in S \backslash Q'_T$ **then**
3:         decrement S.intersection by 1
4:     **else**
5:         decrement S.union by 1
6:     **if** $IN \in S \backslash Q'_T$ **then**
7:         increment S.intersection by 1
8:     **else**
9:         increment S.union by 1
10:     $\mathrm{sim}(S, Q_T) = \dfrac{s.\text{intersection}}{s.\text{union}}.$

Fig. 1. Simple algorithm which updates the similarity values at T

one inverted list $l(x)$ for each element $x \in \Phi$. $l(x)$ stores the IDs of sets which contain an element x and provides fast access to them. However, SIMPLE must access to not only the sets with either IN or OUT but also the sets without IN and OUT in the 4th and 8th lines in Fig. 1. It is difficult to access the latter quickly with the standard inverted lists.

Subsequently, Sect. 4.2 refines SIMPLE wisely, so that the access to the sets without IN and OUT may be eliminated. Then, Sect. 4.3 presents our method which accelerates the refined algorithm with frequency-based inverted lists (FIL). FIL accesses quickly to the sets S satisfying $IN \in S \backslash Q'_T$ or $OUT \in S \backslash Q'_T$.

4.2   Elimination of Access to Sets Without a Specific Element

This subsection shows how to remove the access to sets without IN and OUT from SIMPLE in order to make it more compatible with inverted lists.

In SIMPLE, if S satisfies both $OUT \in S \backslash Q'_T$ and $IN \in S \backslash Q'_T$, S.intersection at T remains unchanged from that at $T-1$, since S.intersection first decrements and then increments. In the same way, if $OUT \notin S \backslash Q'_T$ and $IN \notin S \backslash Q'_T$ at the same time, S.union at T equals that at $T - 1$, since S.union first decrements and then increments. Therefore, the 4 conditional statements in SIMPLE are aggregated into the 2 conditional statements C1 and C2 below.

(C1) If $OUT \in S \backslash Q'_T$ and $IN \notin S \backslash Q'_T$, decrement S.intersection and increment S.union.
(C2) If $IN \in S \backslash Q'_T$ and $OUT \notin S \backslash Q'_T$, increment S.intersection and decrement S.union.

Interestingly, C1 and C2 can be further amended to the new conditional statements N1 and N2 below which completely exclude the access to sets without IN and OUT.

(N1) If $OUT \in S \backslash Q'_T$, decrement S.intersection and increment S.union.
(N2) If $IN \in S \backslash Q'_T$, increment S.intersection and decrements S.union.

Let us confirm that N1 and N2 are equivalent with C1 and C2. Any set S is categorized into one of the following 4 types: (Type 1) $OUT \notin S \backslash Q'_T$ and $IN \notin S \backslash Q'_T$, (Type 2) $OUT \in S \backslash Q'_T$ and $IN \notin S \backslash Q'_T$, (Type 3) $OUT \notin S \backslash Q'_T$ and $IN \in S \backslash Q'_T$ and (Type 4) $OUT \in S \backslash Q'_T$ and $IN \in S \backslash Q'_T$. We show that N1 and N2 behave in the same way as C1 and C2 for all the types.

If S belongs to (Type 1), neither C1 and C2 nor N1 and N2 process S. If S belongs to (Type 2), N1 processes S in the same way as C1. If S belongs to (Type 3), N2 processes S in the same way as C2. Finally, when S belong to (Type 4), N1 decrements S.intersection and increments S.union, whereas N2 increments S.intersection and decrements S.union. In total, N1 and N2 do not change either S.intersection or S.union, which is also the case for C1 and C2.

Thus, N1 and N2 remove the access to sets without IN and OUT.

4.3 Our Algorithm: EA-FIL

Here, we present our final algorithm which executes the modified conditional statements N1 and N2 quickly with frequency-based inverted lists (FIL). Hence, our algorithm is named EA-FIL (Exact Algorithm based on FIL). FIL can discover the sets S such that $OUT \in S \backslash Q'_T$ or $IN \in S \backslash Q'_T$ efficiently.

FIL arranges a list $l(x, \alpha)$ for a combination of $x \in \Phi$ and a natural number α. $l(x, \alpha)$ stores the IDs of sets in the database D which contain α instances of x. FIL is constructed by scanning D only once. Moreover, FIL does not increase the total number of sets in all the lists as compared with the standard inverted lists. We implement FIL as a two-dimensional array of inverted lists indexed by the element ID and the frequency of elements. Alternatively, one may implement FIL with so-called frequency-ordered inverted lists [7].

FIL can find any set $S \in D$ satisfying $IN \in S \backslash Q'_T$ as follows. Consider the situation in which we have Q'_T at hand. First, we investigate the frequency of IN in Q'_T. Let this frequency be β. Note that $IN \in S \backslash Q'_T$ iff there exist strictly more instances of IN in S than in Q'_T. Therefore, we can collect all the sets $\{S \in D | IN \in S \backslash Q'_T\}$ by traversing the group of lists $\{l(IN, \gamma) | \gamma \geq \beta + 1\}$. Significantly, we can obtain β efficiently without scanning the whole Q'_T by maintaining the histogram of alphabets for the query dynamically as follows: Initially, we make the histogram of alphabets for Q_1 by counting each alphabet in Q_1. Then, when the time advances from $T - 1$ to T for $T \geq 2$, we can obtain the histogram for Q'_T only by subtracting the frequency of OUT by one in the histogram for Q_{T-1}. β is immediately obtained by examining this histogram. After this, the histogram for Q_T is computed by increasing the frequency of IN by one in the histogram for Q'_T. FIL can also search a set S satisfying $OUT \in S \backslash Q'_T$ quickly in the same manner.

In Fig. 2, we show our exact algorithm EA-FIL which returns the most similar k sets to Q_T at T. The **IF** statement in the first line avoids the wasteful processing for the trivial case that $IN = OUT$ in which the similarity value never changes for any set in D. When $OUT \in S \backslash Q'_T$ and $IN \in S \backslash Q'_T$, EA-FIL divides the intersection size by the union size twice in the 5th and 9th lines. Though this appears to be redundant, by doing so, we do not have to remember

1: **if** $IN \neq OUT$ **then**
2:     find all the sets S satisfying $OUT \in S \backslash Q'_T$ via FIL.
3:     **for** S s.t. $OUT \in S \backslash Q'_T$ **do**
4:         decrement S.intersection and increment S.union.
5:         $\text{sim}(S, Q_T) = \dfrac{S.\text{intersection}}{S.\text{union}}$.
6:     find all the sets S satisfying $IN \in S \backslash Q'_T$ via FIL.
7:     **for** S s.t. $IN \in S \backslash Q'_T$ **do**
8:         decrement S.intersection and increment S.union.
9:         $\text{sim}(S, Q_T) = \dfrac{S.\text{intersection}}{S.\text{union}}$.
10: Select the top-k sets whose similarity values are the largest.

Fig. 2. Algorithm EA-FIL

whether S has satisfied $OUT \in S \backslash Q'_T$, when the second **FOR** statement starts at the 7th line.

As a summary, at time T, EA-FIL recognizes the sets whose similarity values can change from time $T-1$ efficiently via FIL and processes them only. Moreover, the similarity values at T are derived in $O(1)$ time by updating the old similarity values at $T - 1$, as presented in the 4th and 8th lines in Fig. 2.

5   Experiments

We evaluate our EA-FIL experimentally on synthetic datasets and two real datasets. The experimental platform is a PC with Intel Core i7-4790 CPU@ 3.60 GHz, 8 GB memory and Ubuntu 14.04. For comparison, we also implement (1) the brute-force method BFM which computes the Jaccard similarity n times at each time instant, (2) the previous pruning-based exact algorithm GP in Sect. 3.1 and (3) the previous approximation algorithm MHI in Sect. 3.2. As for MHI, we configure the number of hash functions r to 100. We follow the experimental procedure in the previous work [8] as faithfully as possible.

5.1   Results on Synthetic Datasets

First, we explain how to synthesize the database D. Like [8], after fixing the query size W, we randomly make n sets whose elements are chosen from the set of alphabets Φ with the IBM Quest data generator, where the average set cardinality is set to W. The database generation is controlled by the three parameters (1) W: the query size, (2) $|\Phi|$: the number of alphabet kinds and (3) n: the number of sets in D.

Next, in order to generate the data stream, we first prepare a long sequence of alphabets by randomly concatenating the sets in D whose cardinalities lie between $0.8W$ and $1.2W$. Then, the data stream is simulated by passing one element from the long sequence from the beginning at each time instant.

We evaluate the efficiency of an algorithm by measuring the total processing time to perform the continuous top-k similarity search 1000 times over the time period lasting from $T = 1$ to $T = 1000$, where $k = 10$.

Again like [8], with specifying the parameter configuration that $W = 10$, $|\Phi| = 10000 = 10^4$ and $n = 100000 = 10^5$ as the base case, we vary only one of the three parameters to examine how each parameter affects the efficiency.

Experiment (1): When n is changed

Figure 3(a) shows the total processing time of the four methods, i.e., our EA-FIL, GP, MHI and BFM for various values of $n = \{10^4, 5 \times 10^4, 10^5, 5 \times 10^5, 10^6\}$ under the condition $W = 10$ and $|\Phi| = 10000$. This graph tells that our EA-FIL by far outperforms the previous exact algorithm GP. For example, for $n = 10^6$, EA-FIL spends only 2.2 s, whereas the processing time of GP reaches 34.4 s. Thus, EA-FIL runs more than 15 times faster than GP. Notably, the speed of EA-FIL is comparable to the previous approximation algorithm MHI. Since MHI is an approximation algorithm, it often fails to discover the correct answers. When the number of hash functions $r = 100$, we observed that about 15% of the sets answered by MHI are false positives and not the top-k most similar sets to the query in practice. Though MHI can reduce the number of false positives by increasing r, the processing speed will slow down.

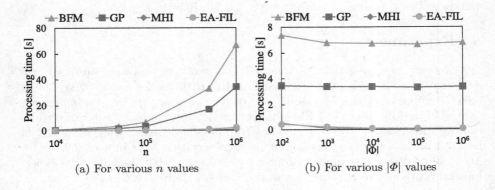

(a) For various n values   (b) For various $|\Phi|$ values

Fig. 3. Processing time for synthetic datasets

Experiment (2): When $|\Phi|$ is changed

With setting W to 10 and n to 10^5, this experiment changes $|\Phi|$ in the range from 10^2 to 10^6. Figure 3(b) displays the total processing time. Also in this experiment, for any $|\Phi|$, EA-FIL is by far superior to GP and runs as fast as MHI. On the other hand, EA-FIL and MHI slightly increase the processing time, as $|\Phi|$ becomes smaller. This feature is interpreted as follows: EA-FIL and MHI rely on the inverted lists. When $|\Phi|$ decreases, we have a smaller number of inverted lists. So that one inverted list maintains more sets, which, in turn, forces the algorithms to process more sets. However, we would emphasize that EA-FIL runs more than 5 times faster than GP, even when $|\Phi|$ is as small as 100.

Experiment (3): When W is changed

We change W in the range $\{10, 50, 100, 500, 1000\}$, while n and $|\Phi|$ are fixed. Figure 4 shows the total processing time for various W values. However, this graph does not display the result for BFM, since BFM was too slow for large W values. Also in this experiment, EA-FIL overwhelms GP and becomes as fast as MHI. Significantly, as W augments, the gap between EA-FIL and GP expands. For example, when $W = 1000$, the processing time reaches 41.4 s for GP, whereas EA-FIL consumes only 1.14 s. Thus, EA-FIL is more than 30 times faster than GP. This result demonstrates the merit of EA-FIL well that always updates the Jaccard similarity values in $O(1)$ time. By contrast, GP has to compute them in $O(W)$ time.

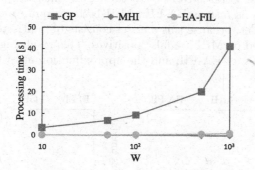

Fig. 4. Processing time for various W values

5.2 Results on Real Datasets

Next, our EA-FIL is evaluated on two real datasets: a Market Basket dataset[1] from an anonymous Belgian retail store and a Click Stream dataset[2]. Both of them were also used in the previous work [8].

The Market Basket dataset is a collection of itemsets each of which presents a set of items purchased by one customer. For the CSPEQ, this dataset serves as a database D and consists of $n = 88162$ sets, where $|\Phi| = 16470$ denotes the number of distinct item kinds. Since the average set cardinality in D becomes 10.3, we set the query size W to 10.

The Click Stream dataset is a collection of sequences of web pages (URLs) visited by some users. By associating each web page with some predefined category such as "news" and "tech", a sequence of web pages visited by a user is represented as a *multiset* of alphabets, where $|\Phi| = 17$ corresponds to the number of categories. For this dataset, the database D consists of $n = 31790$ sets whose average set cardinality equals 13.33. Thus, W is set to 13 for this dataset. The

---

[1] http://fimi.ua.ac.be/data/.

[2] http://www.philippe-fournier-viger.com/spmf/index.php?link=datasets.php.

Click Stream dataset contrasts with the Market Basket dataset, since $|\Phi|$ is very small for the Click Stream dataset.

Given D and the value of W, the data stream is synthesized completely in the same way as in Sect. 5.2. We evaluate an algorithm with the total time to process the continuous top-k similarity search over the time period lasting from $T = 1$ to $T = 1000$. We vary k in the range $\{1, 10, 100, 1000\}$.

Figure 5(a) shows the processing time for the Market Basket dataset. It shows that EA-FIL runs faster than GP by one order of magnitude and as fast as MHI.

Figure 5(b) shows the processing time for the Click Stream dataset. Again, EA-FIL defeats GP in terms of the execution speed. For instance, when $k = 100$, EA-FIL runs 5 times faster than GP. It is magnificent that EA-FIL operates much faster than GP even for this dataset, because this dataset has an extremely small $|\Phi|$ value and works adversely to the algorithms based on inverted lists. Though MHI runs as fast as EA-FIL, MHI behaves quite inaccurately for the Click Stream dataset, because this dataset consists of multisets. More than 60% of the sets answered by MHI are false positives. This result highlights the difference between our exact EA-FIL and the approximation algorithm MHI.

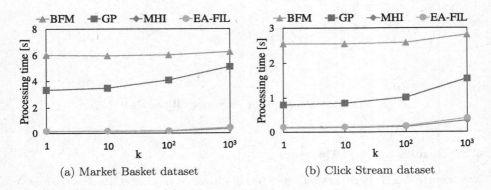

(a) Market Basket dataset   (b) Click Stream dataset

Fig. 5. Processing time for real datasets

6   Related Works

Since the CSPEQ is a new problem, this paper is the second work which studies this problem, as far as we know. Therefore, [8] is the only previous work closely related to our paper. Below we refer to related researches which treat the similarity search on data streams. The similarity search on data streams is logically classified into three types: (Type 1) Only the database changes dynamically (Type 2) Only the query evolves dynamically and (Type 3) Both of the query and the database change. The CSPEQ belongs to (Type 2).

In the literature, (Type 1) has been most intensively researched. Yang et al. [9] studied the continuous top-k search problem, where the query is a static function F which represents a preference of a given user. The purpose of this problem is to maintain the top-k objects which maximize F. Since each object has its own lifetime, we must update the top-k objects, whenever the lifetime of some object in the top-k expires. Rao *et al.* [6] examined the algorithm to process multiple queries efficiently by sharing the function evaluation results among them. Lian *et al.* [5] considered a problem which involves m data streams of real numbers. Here, the database consists of m W-dimensional vectors each of which consists of the latest W real numbers in a single data stream. They realized the approximate similarity search by updating the Locality-Sensitive Hashing [2] functions dynamically. Kontaki *et al.* [4] addressed a similarity search problem of (Type 3), where a feature of a data stream is expressed with its DFT (Discrete Fourier Transform) coefficients.

Datar *et al.* [3] developed a fast algorithm to update the hash values of Min-Hash for evolving sets. This algorithm may serve as a subroutine in MHI.

7   Conclusion

We present a new exact algorithm for the continuous similarity search problem for evolving queries (CSPEQ). Whereas the previous exact algorithm decides not to compute the similarity value for a set S at time T from the magnitude of past similarity value of S, our EA-FIL computes the similarity values only for sets whose similarity values at T change from time $T-1$. At T, our EA-FIL quickly identifies such sets with frequency-based inverted lists (FIL) and derives their similarity values in $O(1)$ time by updating the previous similarity values at $T-1$. Experimentally, EA-FIL runs faster than the previous exact algorithm by one order of magnitude and as fast as the previous approximation algorithm.

Our primary novelty is to show that the CSPEQ can be solved efficiently with inverted lists. Specifically, we designed EA-FIL carefully, so that it may not process the sets without a specific element which are hard to discover quickly via inverted lists.

Acknowledgments. This work was supported by JSPS KAKENHI Grant Number JP15K00148, 2016.

References

1. Broder, A.Z., Charikar, M., Frieze, A.M., Mitzenmacher, M.: Min-wise independent permutations. J. Comput. Syst. Sci. **60**(3), 630–659 (2000)
2. Datar, M., Immorlica, N., Indyk, P., Mirrokni, V.S.: Locality-sensitive hashing scheme based on p-stable distributions. In: Proceedings of ACM SCG 2004, pp. 253–262. ACM (2004)
3. Datar, M., Muthukrishnan, S.: Estimating rarity and similarity over data stream windows. In: Möhring, R., Raman, R. (eds.) ESA 2002. LNCS, vol. 2461, pp. 323–335. Springer, Heidelberg (2002). doi:10.1007/3-540-45749-6_31
4. Kontaki, M., Papadopoulos, A.N., Manolopoulos, Y.: Adaptive similarity search in streaming time series with sliding windows. Data Knowl. Eng. **63**(2), 478–502 (2007)

5. Lian, X., Chen, L., Wang, B.: Approximate similarity search over multiple stream time series. In: Kotagiri, R., Krishna, P.R., Mohania, M., Nantajeewarawat, E. (eds.) DASFAA 2007. LNCS, vol. 4443, pp. 962–968. Springer, Heidelberg (2007). doi:10. 1007/978-3-540-71703-4_86
6. Rao, W., Chen, L., Chen, S., Tarkoma, S.: Evaluating continuous top-k queries over document streams. World Wide Web 17(1), 59–83 (2014)
7. U, L.H., Zhang, J., Mouratidis, K., Li, Y.: Continuous top-k monitoring on document streams. IEEE Trans. Knowl. Data Eng. 29(5), 991–1003 (2017)
8. Xu, X., Gao, C., Pei, J., Wang, K., Al-Barakati, A.: Continuous similarity search for evolving queries. Knowl. Inf. Syst. 48(3), 649–678 (2016)
9. Yang, D., Shastri, A., Rundensteiner, E.A., Ward, M.O.: An optimal strategy for monitoring top-k queries in streaming windows. In: Proceedings of the 14th International Conference on Extending Database Technology, pp. 57–68. ACM (2011)

Personalization and Recommendation

A Collaborative Neural Model for Rating Prediction by Leveraging User Reviews and Product Images

Wenwen Ye[1], Yongfeng Zhang[2], Wayne Xin Zhao[3(✉)], Xu Chen[1], and Zheng Qin[1]

[1] School of Software, Tsinghua University, Beijing, China
[2] College of Information and Computer Sciences, University of Massachusetts Amherst, Amherst, USA
[3] School of Information, Renmin University of China, Beijing, China
batmanfly@gmail.com

Abstract. Product images and user reviews are two types of important side information to improve recommender systems. Product images capture users' appearance preference, while user reviews reflect customers' opinions on product properties that might not be directly visible. They can complement each other to jointly improve the recommendation accuracy. In this paper, we present a novel collaborative neural model for rating prediction by jointly utilizing user reviews and product images. First, product images are leveraged to enhance the item representation. Furthermore, in order to utilize user reviews, we couple the processes of rating prediction and review generation via a deep neural network. Similar to the multi-task learning, the extracted hidden features from the neural network are shared to predict the rating using the softmax function and generate the review content using LSTM-based model respectively. To our knowledge, it is the first time that both product images and user reviews are jointly utilized in a unified neural network for rating prediction, which can combine the benefits from both kinds of information. Extensive experiments on four real-world datasets demonstrate the superiority of our proposed model over several competitive baselines.

1 Introduction

Nowadays, recommender systems have been widely used in various online services, such as e-commerce, news-reading and video-sharing websites. Traditional methods mainly capture the interactions between users and items, *e.g.,* factorizing the user-item rating matrix [14]. Recently, with the ever increasing of user-generated content, multiple types of side information have been utilized to improve the recommendation accuracy. Among these side information, product images and user reviews have received much research attention.

Intuitively, product images can directly reflect users' appearance preference (*e.g.,* clothing styles and phone looks), which are usually seldom (or even hard)

© Springer International Publishing AG 2017
W.-K. Sung et al. (Eds.): AIRS 2017, LNCS 10648, pp. 99–111, 2017.
https://doi.org/10.1007/978-3-319-70145-5_8

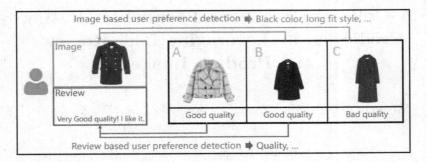

Fig. 1. An illustrative example on the complementary effect of visual and textual features on recommender systems.

to be described in words, while user reviews can uncover the customers' favored characters that might be invisible from the product images (*e.g.,* clothing quality and phone weight), they can complement each other for better understanding users' interests. To see this, we present an illustrative example in Fig. 1. A user has assigned a high rating and posted a short review on an item: *"Very good quality! I like it.".* On one hand, from the product image, we can clearly identify the visual characteristics that the user prefers, *e.g.,* the color and the style. On the other hand, as a complementary signal, the user review can further reveal the other important aspects considered by the user, *i.e.,* quality. In this example, with the help of visual features, B and C could be selected as candidate recommendations to the user, since they are similar to the purchased product by her in terms of color and style. Furthermore, by considering review information, C will be filtered out due to bad experiences on *clothing quality* indicated in historical reviews, which is an important factor to consider for the current user.

Existing works have demonstrated the effectiveness of using either user reviews or product images [9,15,17] for rating prediction. However, few studies have investigated the effect of their integration for more accurate recommendations. Hence, we would like to study two research questions: (1) whether it is possible to jointly utilize user reviews and product images in a unified model; (2) how such an integration improves the performance compared with traditional methods using a single kind of information.

A major challenge to answer the two questions is how to properly and effectively combine heterogeneous information (*i.e.,* user review and product image) together. Traditional methods [14,20] would be less effective when faced with multiple kinds of heterogeneous information due to the limitation of simple linear structures [10]. Fortunately, the rapid development of deep learning techniques sheds light on this problem because of its superiority in the field of multi-modal fusing [13,19,21,24].

In this paper, we present a novel collaborative **N**eural **R**ating **P**rediction model by jointly modeling the **T**extual features collected from user reviews and the **V**isual features extracted from product images (called **NRPTV**). We develop the model in a deep learning framework. Our model is built on the core components, *i.e.*, user and item representations, which encodes useful information from users and items. To integrate visual features, we combine the item latent factor (obtained by using a lookup layer) with the transformed visual features (derived from item images) as the image-enhanced item representation. The derived user and item representations are subsequently fed into a Multi-Layer Perceptron (MLP) as input for rating prediction. Furthermore, in order to utilize user reviews, we couple the processes of rating prediction and review generation via a MLP component. Similar to the multi-task learning, the extracted hidden features from the MLP component are shared to predict the rating using the softmax function and generate the review content using a LSTM-based model respectively. In this way, our model can utilize both textual and visual features for recommender systems, and combine the benefits from both kinds of features.

To the best of our knowledge, it is the first time that both product images and user reviews are jointly characterized in a unified neural network model for rating prediction. Extensive experiments on four real-world datasets demonstrate the superiority of our proposed model over several competitive baselines. Our experiment results also show that using a combination of both types of features leads to a substantial performance improvement compared to that using only either type.

In the rest of the paper, we first review the related work in Sect. 2, and present the proposed model in Sect. 3. Section 4 gives the experimental results and analysis, and Sect. 5 concludes the paper.

2   Related Work

Recommender systems have attracted much attention from the research community and industry [2]. We mainly review two highly related research directions.

Review-Based Recommendation. Many efforts have been made to incorporate user reviews into traditional recommendation algorithms [29]. The major assumption is that review contain important textual features which are potentially improve the recommendation performance. Typical methods include correlating review topics with the latent factors in matrix factorization [15, 25], feature-level information utilization [3, 31], and the distributed semantic modeling [28]. A problem with these studies is that they usually make the bag-of-words assumption, and sequential word order has been ignored.

Image-Based Recommendation. In recent years, visual features have been leveraged by recommender systems [4, 9, 17, 30], which aim to capture important appearance characteristics of the items. Specially, visual features have been incorporated into the Bayesian Personalized Ranking framework [9] and sequential prediction task [6]. Furthermore, visual features have been used to better find visually complementary items to a query image [17].

Although review and image data are important and complementary to recommender systems, to our knowledge, few studies can jointly utilize both user review and product image. Our work take the initiative to develop a collaborative neural model leveraging both visual and textual features for rating prediction. We are also aware of the recent progress of deep learning techniques on recommender systems [5, 10, 25, 26, 32]. They seldom consider the incorporation of review and image. As a comparison, our focus is to borrow the benefits of deep learning models and perform effective heterogeneous data fusion and utilization for rating prediction.

3   A Collaborative Neural Model for Rating Prediction

In this section, we present the preliminaries and the proposed collaborative neural models for rating prediction.

3.1   Preliminaries

Formally, let u and i denote a user and an item respectively, and $r_{u,i}$ denote the rating of user u on item i. As usual, the rating values are discrete and range from 1 to R, *e.g.*, the five-star rating mechanism on Amazon ($R = 5$). We assume the corresponding images for item i are available, denoted by a vector $\boldsymbol{v}_i$. Each rating $r_{u,i}$ is associated with a user review, denoted by a vector $\boldsymbol{w}_{u,i}$, in which $w_{u,i,k}$ denotes the k-th word token in the review. Given an observed rating dataset $\mathcal{D} = \{\langle u, i, r_{u,i}, \boldsymbol{w}_{u,i}, \boldsymbol{v}_i \rangle\}$, the *rating prediction* task aims to predict the ratings of a user on the un-rated items.

A typical approach to rating prediction is to factorize the interactions between users and items by learning user and item representations. In standard matrix factorization (MF), it approximately estimates the missing entry $r_{u,i}$ by the inner product between user latent representation $\boldsymbol{x}_u$ and item latent representation $\boldsymbol{y}_i$, *i.e.*, $\hat{r}_{u,i} = \boldsymbol{x}_u^\top \cdot \boldsymbol{y}_i$. Although such a general approach is widely used, it adopts the linear matrix factorization, and may not be effective and flexible to incorporate multiple types of heterogeneous information. Inspired by the recent progress in the multi-modality deep learning, next we build our approach in a deep learning framework.

3.2   The Base Model: Image-Enhanced Rating Prediction

Following the classic MF approach, our base model also keeps user and item representations. Furthermore, we assume that both kinds of representations not only characterize the user-item interactions, but also can encode other useful side information for rating prediction. Formally, let $\tilde{\boldsymbol{x}}_u$ and $\tilde{\boldsymbol{y}}_i$ denote the user and item representations in our model, both of which are a K-dimensional vector, *i.e.*, $\tilde{\boldsymbol{x}}_u, \tilde{\boldsymbol{y}}_i \in \mathbb{R}^K$. The focus of this section is to incorporate image information into the item representation and develop a neural model for rating prediction.

User and Item Embeddings as Representations. First, with the help of the look-up layer, the one-hot input of either user or item IDs are first projected

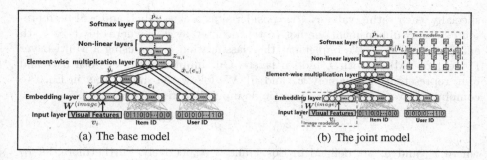

Fig. 2. Our proposed base model (left) and joint model (right) for rating prediction. Boxes with dotted circles represent embedding or hidden vectors. The gray, green and orange boxes correspond to v_i, e_u and e_i respectively. (Color figure online)

into low-dimensional embeddings, which are similar to the latent factors in the context of matrix factorization. Let $e_u(\in \mathbb{R}^K)$ and $e_i(\in \mathbb{R}^K)$ denote the user and item embeddings, we then set the basic user and item representations as the corresponding embeddings by:

$$\tilde{x}_u = e_u, \tag{1}$$
$$\tilde{y}_i = e_i. \tag{2}$$

Imaged-Enhanced Item Representation. As shown in Fig. 1, product images encode appearance characteristics which are potentially useful to improve recommender systems. Now, we study how to incorporate product images in our model. Recall that each item i is associated with a visual feature vector, denoted by v_i. To set v_i, we use a pre-trained approach to generate visual features from raw product images using the deep learning framework of CAFFE [12]. Following [9], we adopt the CAFFE reference model with five convolutional layers followed by three fully-connected layers that has been pre-trained on 1.2 million IMAGENET images. For item i, the second fully-connected layer is taken as the visual feature vector v_i, which is a feature vector of length 4096. To combine v_i with e_i, we map v_i into a K-dimensional vector $\tilde{v}_i$ using a linear transformation, i.e., $\tilde{v}_i = W^{(image)} \cdot v_i$, where $W^{(image)} \in \mathbb{R}^{K \times 4096}$ is the transformation matrix for images. Once $\tilde{v}_i$ and e_i have the same dimensionality, we further preform the element-wise vector product to derive the new item representation,

$$\tilde{y}_i = \tilde{v}_i \odot e_i. \tag{3}$$

Note that there can be alternative ways to combine $\tilde{v}_i$ with e_i, i.e., the vector concatenation. In our experiments, our current choice leads to a good performance and we adopt it as the combination method.

Rating Prediction Through a MLP Classifier. Recall that our rating values have totally R choices. Hence, we adopt a classification approach to solve the rating prediction task, which has been shown to be effective in [23,33]. More specially, each rating value $r_{u,i}$ is considered as a class label, and will be represented as a R-dimensional one-hot vector, denoted by $\boldsymbol{p}_{u,i}$, where only the $r_{u,i}$-th entry is equal to 1. We implement the classification model using a Multi-Layer Perceptron (MLP) with L hidden layers. Our input consists of both user and item representations, namely $\tilde{\boldsymbol{x}}_u$ and $\tilde{\boldsymbol{y}}_i$. We follow the similar way in Eq. 3 to combine two representation into a single input feature vector $\boldsymbol{z}_{u,i} \in \mathbb{R}^K$

$$\boldsymbol{z}_{u,i} = \tilde{\boldsymbol{x}}_u \odot \tilde{\boldsymbol{y}}_i, \tag{4}$$

where $\tilde{\boldsymbol{x}}_u$ and $\tilde{\boldsymbol{y}}_i$ are defined in Eqs. 1 and 3 respectively. Furthermore, let $\boldsymbol{h}_l$ denote the corresponding output of the l-th hidden layer, which is derived on top of the $(l-1)$-th hidden layer for $l \in \{1, 2, ...L\}$

$$\boldsymbol{h}_l = f(\boldsymbol{h}_{l-1}), \tag{5}$$

where $f(\cdot)$ is an non-linear activation function implemented by the Rectifier Linear Unit (ReLU) in our model because ReLU is usually more resistable to overfitting and works well in practice [7]. To feed the input of the MLP classifier, we set $\boldsymbol{h}_0 = \boldsymbol{z}_{u,i}$. Finally, the last softmax layer takes in $\boldsymbol{h}_L$ to produce a probability distribution for R classes, denoted by $\hat{\boldsymbol{p}}_{u,i}$. The loss function computes the cross-entropy loss between the ground truth (i.e., $\boldsymbol{p}_{u,i}$) and the predicted results (i.e., $\hat{\boldsymbol{p}}_{u,i}$):

$$\mathcal{L}_{base} \doteq \sum_{\langle u,i \rangle \in \mathcal{D}} \mathrm{CrossEnt}(\boldsymbol{p}_{u,i}, \hat{\boldsymbol{p}}_{u,i}), \tag{6}$$

$$= \sum_{\langle u,i \rangle \in \mathcal{D}} \sum_{r=1}^{R} -p_{u,i,r} \cdot \log \hat{p}_{u,i,r}.$$

We present a model sketch in Fig. 2(a). As we can see, the model has incorporated the visual features in the item representation. The combined user and item representations are fed into a MLP classifier with L non-linear layers. We adopt the non-linear transformation since visual features may not be directly ready in a form for rating prediction. Deep neural models could be effective to transform the visual information into a better representation for rating prediction.

3.3 The Joint Model: Integrating User Review with Product Image for Rating Prediction

In the above, visual features have been fused into rating prediction model via the improved item representation. Now, we study how to integrate user reviews into the base model. Following Sect. 3.2, we can take a similar approach to incorporate review information, in which the textual features would be used as input to enhance either the user or item representations. However, such a straightforward approach may be practically infeasible because: (1) given a user, her review information on some product may not be available for rating prediction model, since the reviewing behavior usually occur after purchase behavior; and (2) user reviews and product images represent two kinds of heterogeneous side information, it is likely to perform poorly by simply fusing two kinds of information.

Overview of the Model. To address this difficulty, our idea is to treat the review content as another kind of output signal besides the ratings. We do not modify the bottom neural architecture for user and item representations in the base model. Instead, we take the output of the last non-linear layer (*i.e.*, h_L) in the MLP component, and generate review contents (*i.e.*, $w_{u,i}$) based on it. Since h_L was previously passed into a softmax layer for rating prediction, our current approach couples the two processes: rating prediction and review generation. In essence, the idea is similar to the multi-task or multi-modality learning [1]. Following this idea, the next problem is how to model the review generation.

LSTM-Based Review Modeling. Most of the existing review-based recommendation models make the bag-of-words assumption [15,22], and they ignore the effect of sequential word order on the semantics of review text. Consider two sample reviews: "*The screen is good, while the battery is unsatisfactory*" and "*The screen is unsatisfactory, while the battery is good*". Although they consist of the same words, they convey totally different semantics. Hence, we have to consider the sequential order in review generation. To capture the word sequential information, we adopt the long short term memory (LSTM) [11] network, which has been successfully applied to a number of sequence text modeling tasks [24,27]. Formally, let V be the vocabulary size, and $w_{u,i} = \{w_{u,i,0}, ..., w_{u,i,k}, ..., w_{u,i,n_{u,i}-1}\}$ denote the review published by user u on item i, where $n_{u,i}$ is the review length and $w_{u,i,k}$ is the $(k+1)$-th token in the review. The LSTM model generates the review content in a sequential way as follows:

$$s_k = \text{LSTM}(s_{k-1}, w_{u,i,k-1}, \Phi), \tag{7}$$
$$p(w_{u,i,k}|w_{u,i,<k}, \Phi) = \text{softmax}_{w_{u,i,k}}(s_k, \Phi), \tag{8}$$

where $1 \leq k \leq n_{u,i} - 1$, $w_{u,i,<k}$ denote the preceding k words, the s_k is the state vector for the k-th step (*i.e.*, the k-th word), LSTM$(\cdot)$ is the standard LSTM unit [11], softmax$(\cdot)$ is the softmax function which transforms the hidden state into a V-dimensional word generation probability distributions, and Φ denote the set of all the necessary parameters for text generation.

The Joint Optimization Model. The above formulation presents the review text generation independent from the rating prediction. Next, we couple these two parts via the shared hidden layer in the MLP component. We set the initial state vector to the last hidden layer in the MLP as below

$$s_0 = h_L, \tag{9}$$

where h_L is defined by Eq. 5 in Sect. 3.2 and the subscript of "0" indicates the zero state of LSTM. Finally, the overall loss function is given below

$$
\begin{aligned}
\mathcal{L}_{joint} = & \alpha \sum_{\langle u,i \rangle \in \mathcal{D}} \sum_{k=1}^{n_{u,i}-1} -\log p(w_{u,i,k}|\boldsymbol{w}_{u,i,<k}, \Phi) \\
& + (1-\alpha) \sum_{\langle u,i \rangle \in \mathcal{D}} \text{CrossEnt}(\boldsymbol{p}_{u,i}, \hat{\boldsymbol{p}}_{u,i}),
\end{aligned}
\tag{10}
$$

where $\text{CrossEnt}(\boldsymbol{p}_{u,i}, \hat{\boldsymbol{p}}_{u,i})$ is the cross-entropy defined in Eq. 6, $p(w_{u,i,k}|\boldsymbol{w}_{u,i,<k}; \Phi)$ is the word generation probability defined in Eq. 8, and α ($0 \leq \alpha < 1$) is a tuning parameter that balances the effects of the two parts. When $\alpha = 0$, the model becomes the base model in Sect. 3.2. In our joint model, the visual features are integrated via the item representation, and the textual features are integrated in the output signals. Note that although we describe the two parts separately, our approach links both parts in a unified optimization model. All the parameters can be jointly learned by optimizing the loss function Eq. 10 using the stochastic gradient descent (SGD) method.

We denote the model by **NRPTV**, *i.e.,* a collaborative **N**eural **R**ating **P**rediction model based on **T**extual features and **V**isual features. We present the overview of the final model in Fig. 2(b). By comparing Fig. 2(a) and (b), we can see that the textual features have been integrated into the base model in a similar way as multi-task learning. The purpose of the MLP component is to enhance the capacity to integrate heterogeneous information and improve the prediction performance.

4   Experiments

In this section, we present the experiments. We begin by introducing the experimental setup, and then report and analyze the experimental results.

4.1   Datasets

We use four Amazon datasets shared in [16], which are from four diverse product categories. The statistics of these datasets are shown in Table 1.

Table 1. Basic statistics of the datasets.

Datasets	#Users	#Items	#Ratings	$\frac{\#Reviews}{\#Users}$	Density
Music	1,492	900	7,931	5.55	0.59%
Patio	1,686	962	11,740	6.96	0.72%
Auto	2,928	1,835	18,308	6.25	0.34%
Clothing	39,387	23,033	278,677	7.07	0.03%

Methods to Compare. To demonstrate the effectiveness of our model, we consider the following methods for performance comparison:

- **PMF** [18]: PMF model represents the classic rating prediction approach, which only utilizes the user-item rating matrix.
- **HFT** [15]: HFT model aligns topics from topic models with latent factors in matrix factorization. It is a commonly used review-based rating prediction baseline.
- **modified VBPR (mVBPR)** [9]: VBPR is a pioneering work which incorporates product images into top-N recommendation. To adapt VBPR to rating prediction, we modify the original ranking loss to the least square loss.
- **modified NeuMF (mNeuMF)** [10]: NeuMF is a recently proposed neural network model for top-N recommendation. To adapt NeuMF to rating prediction, we modify the original ranking loss to the cross-entropy loss as our model.
- **NRPTV**: NRPTV is our proposed model which jointly utilizes visual and textual information.

4.2   Parameter Settings

Our models are implemented using the library TENSORFLOW. The model parameters are randomly initialized according to the uniform distribution. The learning rate of SGD is determined by grid searching in the set $\{1, 0.1, 0.01, 0.001, 0.0001\}$. We set the number of non-linear layers in the MLP component to 3, and the dimensionality are set to $\{40, 20, 5\}$ to form a tower structure [8]. The tuning parameter α is first set to 0.1 empirically, and will be analyzed in detail in the following experiment. For fair comparison, we set all the biases as 0 in the baseline models. In our experiments, we randomly split the full dataset into training and test sets with a split ratio of 7:3. To evaluate the performance of the comparison methods, RMSE (Root of the Mean Square Error) is adopted as the evaluation metric.

Results and Analysis. In this section, we present the experimental results and analysis on the task of rating prediction. The RMSE results of different methods are reported in Table 2. From Table 2, we can make the following observations:

- Overall, a dimensionality of 100 works well for all the methods. A smaller dimensionality may not be able to achieve powerful predictability, while a larger dimensionality tends to overfit on the training data.
- mNeuMF is better than PMF because it can incorporate more powerful non-linear transformation, which may lead to a better performance.
- With additional side information, either image or review, both mVBPR (+*image*) and HFT (+*review*) further improve substantially over mNeuMF and PMF in most cases. This finding indicates that side information is important to consider in rating prediction.

Table 2. Performance comparison of different methods using RMSE (smaller is better). K denotes the dimensionality.

Datasets	K	PMF	mNeuMF	mVBPR	HFT	NRPTV
Music	50	1.076	1.075	1.069	1.067	**1.063**
	100	1.075	1.072	1.066	1.064	**1.061**
	150	1.077	1.074	1.071	1.068	**1.064**
	200	1.078	1.072	1.068	1.067	**1.066**
Patio	50	1.242	1.240	1.237	1.216	**1.211**
	100	1.236	1.234	1.231	1.215	**1.209**
	150	1.244	1.241	1.239	1.224	**1.213**
	200	1.245	1.241	1.238	1.227	**1.216**
Auto	50	1.179	1.174	1.171	1.169	**1.159**
	100	1.172	1.171	1.170	1.168	**1.157**
	150	1.183	1.181	1.173	1.171	**1.163**
	200	1.189	1.184	1.176	1.173	**1.167**
Clothing	50	1.390	1.386	1.381	1.380	**1.374**
	100	1.389	1.382	1.377	1.379	**1.373**
	150	1.394	1.387	1.382	1.385	**1.376**
	200	1.399	1.392	1.382	1.388	**1.380**

– Our proposed model NRPTV ($+image + review$) is consistently better than all the baselines on the four datasets with four different dimensionalities. This is because: NRPTV jointly utilizes images and reviews, and it also combines the benefits from deep learning.

4.3   Detailed Analysis of Our Model

In the above, we have shown the effectiveness of our model NRPTV. Now, we carry out more detailed analysis for NRPTV in order to analyze the individual effect of different components or parameters on the performance. At each time, we only check one component or parameter, while the rest will be fixed to the optimal settings. In what follows, we fix the dimensionality (*i.e.*, K) as 100, since Table 2 has shown that the dimensionality of 100 gives good performance.

Influence of the tuning parameter α. An important parameter to tune is α in Eq. 10. We vary it from 0.1 to 0.9 with a gap of 0.1. We present the tuning results for α in Fig. 3(b). On the *Music* dataset, the performance achieves the best when $\alpha = 0.4$, while on the *Auto* dataset, the performance achieves the best when $\alpha = 0.6$. Due to space limit, we only report the results on the two datasets. For other datasets, we have also found that $\alpha \in [0.4, 0.6]$ usually gives the best performance. α controls the importance of the review generation component when adding to the base model. The observations indicate we should make a

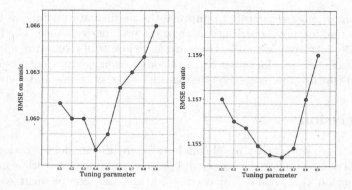

Fig. 3. The influence of tuning parameter α.

suitable balance, neither too large nor too small, between the base model and added review model component.

Influence of the Image and Review Modeling Components. As shown in Fig. 2(b), our model jointly utilizes both the image and review in a unified model. We now examine how each component affects the prediction performance. To examine it, we implement two variants based on the full model, which remove either the review or image components respectively, called $NRPT$ (+*text*) and $NRPV$ (+*image*). Then, we compare the performance of NRPTV, NRPT and NRPV, and report the RMSE results on four datasets in Fig. 4. It can be observed that NRPTV is consistently better than both NRPT and NRPV, which indicates that both components are important to rating prediction. Another interesting finding is that the two variants NRPT and NRPV alternatively perform better than each other. For example, on the *Clothing* dataset, visual features seem

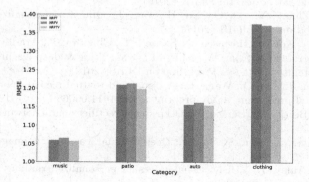

Fig. 4. Performance comparison among NRPT, NRPV and NRPTV based on the RMSE results. NRPT includes the image modeling component, NRPV includes the review modeling component, and NRPTV includes both components.

to play an more important role to improve the performance, while on the *Auto* dataset, textual features contribute more to the final performance. The finding is actually quite intuitive. Visual features are more powerful to capture appearance characteristics such as color and style. As a comparison, textual features are more powerful to capture the usage experiences such as easy to use and convenience.

5   Conclusion

In this paper, we proposed a novel collaborative neural model for rating prediction by jointly modeling user reviews and product images. Our work is motivated by the intuition that visual and textual features can complement each other to improve recommendation accuracy. Extensive experiments demonstrate the effectiveness of our proposed model and the importance to combine both kinds of side information. Our work has made the attempt to characterize heterogeneous side information using deep neural models in recommender systems. As future work, we will consider developing a general model which can integrate more kinds of side information. We will also study how to improve recommendation interpretability using these side information.

Acknowledgment. Xin Zhao was partially supported by the National Natural Science Foundation of China under grant 61502502 and the Beijing Natural Science Foundation under grant 4162032.

References

1. Argyriou, A., Evgeniou, T., Pontil, M.: Multi-task feature learning. In: NIPS (2007)
2. Bobadilla, J., Ortega, F., Hernando, A., Gutiérrez, A.: Recommender systems survey. Knowl. Based Syst. **46**, 109–132 (2013)
3. Chen, X., Qin, Z., Zhang, Y., Xu, T.: Learning to rank features for recommendation over multiple categories. In: SIGIR (2016)
4. Chen, X., Zhang, Y., Ai, Q., Xu, H., Yan, J., Qin, Z.: Personalized key frame recommendation. In: SIGIR (2017)
5. Cheng, H.T., Koc, L., Harmsen, J., Shaked, T., Chandra, T., Aradhye, H., Anderson, G., Corrado, G., Chai, W., Ispir, M., et al.: Wide & deep learning for recommender systems. In: Recsys Workshop on DLRS (2016)
6. Cui, Q., Wu, S., Liu, Q., Wang, L.: A visual and textual recurrent neural network for sequential prediction. arXiv preprint arXiv:1611.06668 (2016)
7. Glorot, X., Bordes, A., Bengio, Y.: Deep sparse rectifier neural networks. In: Aistats (2011)
8. He, K., Zhang, X., Ren, S., Sun, J.: Deep residual learning for image recognition. In: CVPR (2016)
9. He, R., McAuley, J.: VBPR: visual Bayesian personalized ranking from implicit feedback. In: AAAI (2016)
10. He, X., Liao, L., Zhang, H., Nie, L., Hu, X., Chua, T.S.: Neural collaborative filtering. In: WWW (2017)
11. Hochreiter, S., Schmidhuber, J.: Long short-term memory. Neural Comput. **9**(8), 1735–1780 (1997)

12. Jia, Y., Shelhamer, E., Donahue, J., Karayev, S., Long, J., Girshick, R., Guadar-rama, S., Darrell, T.: Caffe: convolutional architecture for fast feature embedding. In: MM (2014)
13. Kiros, R., Salakhutdinov, R., Zemel, R.S.: Multimodal neural language models. In: ICML (2014)
14. Koren, Y., Bell, R., Volinsky, C., et al.: Matrix factorization techniques for recommender systems. Computer **42**(8) (2009)
15. McAuley, J., Leskovec, J.: Hidden factors and hidden topics: understanding rating dimensions with review text. In: Recsys (2013)
16. McAuley, J., Pandey, R., Leskovec, J.: Inferring networks of substitutable and complementary products. In: KDD (2015)
17. McAuley, J., Targett, C., Shi, Q., Van Den Hengel, A.: Image-based recommendations on styles and substitutes. In: SIGIR (2015)
18. Mnih, A., Salakhutdinov, R.: Probabilistic matrix factorization. In: NIPS (2007)
19. Ngiam, J., Khosla, A., Kim, M., Nam, J., Lee, H., Ng, A.Y.: Multimodal deep learning. In: ICML (2011)
20. Rendle, S., Freudenthaler, C., Gantner, Z., Schmidt-Thieme, L.: BPR: Bayesian personalized ranking from implicit feedback. In: UAI (2009)
21. Srivastava, N., Salakhutdinov, R.R.: Multimodal learning with deep Boltzmann machines. In: NIPS (2012)
22. Tan, Y., Zhang, M., Liu, Y., Ma, S.: Rating-boosted latent topics: understanding users and items with ratings and reviews. In: IJCAI (2016)
23. Tang, D., Qin, B., Liu, T., Yang, Y.: User modeling with neural network for review rating prediction. In: IJCAI (2015)
24. Vinyals, O., Toshev, A., Bengio, S., Erhan, D.: Show and tell: a neural image caption generator. In: CVPR (2015)
25. Wang, H., Wang, N., Yeung, D.Y.: Collaborative deep learning for recommender systems. In: SIGKDD (2015)
26. Wang, H., Xingjian, S., Yeung, D.Y.: Collaborative recurrent autoencoder: recommend while learning to fill in the blanks. In: NIPS (2016)
27. Xu, K., Ba, J., Kiros, R., Cho, K., Courville, A.C., Salakhutdinov, R., Zemel, R.S., Bengio, Y.: Show, attend and tell: neural image caption generation with visual attention. In: ICML (2015)
28. Zhang, W., Yuan, Q., Han, J., Wang, J.: Collaborative multi-level embedding learning from reviews for rating prediction. In: IJCAI (2016)
29. Zhang, Y.: Explainable recommendation: theory and applications. arXiv preprint arXiv:1708.06409 (2017)
30. Zhang, Y., Ai, Q., Chen, X., Croft, W.: Joint representation learning for top-n recommendation with heterogeneous information sources. In: CIKM (2017)
31. Zhang, Y., Lai, G., Zhang, M., Zhang, Y., Liu, Y., Ma, S.: Explicit factor models for explainable recommendation based on phrase-level sentiment analysis. In: SIGIR (2014)
32. Zhao, W.X., Li, S., He, Y., Chang, E.Y., Wen, J.R., Li, X.: Connecting social media to E-commerce: cold-start product recommendation using microblogging information. TKDE **28**, 1147–1159 (2016)
33. Zheng, Y., Tang, B., Ding, W., Zhou, H.: A neural autoregressive approach to collaborative filtering. In: ICML (2016)

Wide & Deep Learning in Job Recommendation: An Empirical Study

Shaoyun Shi, Min Zhang$^{(\boxtimes)}$, Hongyu Lu, Yiqun Liu, and Shaopin Ma

Tsinghua National Laboratory for Information Science and Technology,
Department of Computer Science and Technology,
Tsinghua University, Beijing 100084, China
{shisy17,luhy16}@mails.tsinghua.edu.cn,
{z-m,yiqunliu,msp}@tsinghua.edu.cn

Abstract. Recommender systems have become more and more popular in recent years. Collaborative Filtering and Content-Based methods are widely used for a long time. Recently, some researchers introduced deep learning algorithms into recommender system. In this paper, we try to answer some questions about a novel recommender model, Wide & Deep Learning. Firstly, how should we select and feed in features? Secondly, how does Wide & Deep Learning work? Thirdly, how to joint-train the two parts of the network? Finally, how to conduct online training with new data? For all of these, we focus on the job recommendation task, which often suffers from the cold-start problem. The experiments give us the answers of these questions.

Keywords: Recommender system · Wide & Deep Learning · Job recommendation

1 Introduction

With the continuous development of the Internet, information explosion has become a great challenge that people are faced with [1]. How to get the information we need is a big problem in such a situation. As a result, recommender systems are designed to help deal with this problem. Collaborative Filtering and Content-Based methods are widely used for a long time and many studies are based on them. On the other hand, deep learning methods [2] have achieved remarkable results in many fields like image recognition [3] and natural language processing [4], which have also been introduced into recommender systems recently to solve some problems, such as the cold-start problem.

Cold-start is a serious problem in recommendation [5]. There are new items and users coming to the system every day and we do not have any historical information of them. For example, the job recommendation task suffers from the item cold-start problem. There are new jobs published on the website every

This work is supported by Natural Science Foundation of China (Grant No. 61532011, 61672311) and National Key Basic Research Program (2015CB358700).

W.-K. Sung et al. (Eds.): AIRS 2017, LNCS 10648, pp. 112–124, 2017.
https://doi.org/10.1007/978-3-319-70145-5_9

day, which makes it necessary to update the recommendation model. Although traditional Collaborative Filtering methods get remarkable performance, they are weak on solving the cold-start problem because it is difficult for them to update model when fresh users and items come. To better take advantage of users' and items' profile and other external information, deep learning is taken into consideration. Most combinations are implemented by combining deep concepts with traditional methods, such as optimizing the loss of user/item vectors in matrix decompositions by far.

However, there are also new models proposed, such as Wide & Deep Learning [6], which is a novel Content-Based method proposed by Google. Researchers have shown that deep learning has a strong ability to generalize, besides, traditional linear models have a great ability of "memorizing". Both of the two properties are essential in recommendations. Based on this idea, Google proposed Wide & Deep Learning, combining deep neural network with the linear model. It can be regarded as a Content-Based method because the input features are mainly about users' and items' profile and the ratings or other kinds of interaction history are used to supervise the model training.

However, there are still many uncertain questions in applying Wide & Deep. Different kinds of machine learning methods have different abilities to process different kinds of features and data. So what kinds of features should we feed into the Wide & Deep network? Is feature selection necessary? And questions about how Wide & Deep works, such as which part is more important, still have no answer. What's more, due to the phenomenon that the two parts' learning speed is not the same, we need to ensure that the two parts are both well-trained by effective joint-training strategy. Besides, in real-world recommender systems, because of the new data coming every day, it is important to update the model efficiently. We did experiments on all of these questions and got some insights on how to use the Wide & Deep neural network.

The contributions of our work are listed as following:

- We summarized several questions about Wide & Deep which have not been answered by far, while they are important in applying this algorithm.
- Several analysis methods, e.g.: Layer-wise Relevance Propagation (LRP), and Experiments are conducted in job recommendation task to seek the answers to the questions.
- We applied Wide & Deep Learning to job recommendation and achieved good performance.

2   Related Work

2.1   Traditional Methods

The two traditional recommendation techniques are Content-Based and Collaborative Filtering.

Many outstanding Collaborative Filtering methods are based on matrix factorization, such as WRMF [7], ExpoMF [8], MMMF [9] and BPR-MF [10]. Given an interaction matrix $R = (R_{ij})_{m \times n} \in \mathfrak{R}_+^{m \times n}$ of m users and n items, the goal is to get user matrix $U_{m \times k}$ and item matrix $V_{k \times n}$. $R = UV$, where k is the dimension of latent vector. Each row of U represents a user vector u_i and each column of V represents an item vector v_j. We can use u_i and v_j's dot product as the rating prediction about how user i likes item j. But these methods can not deal with users or items with no historical information and it is really hard to add new users and items unless you re-train the model.

On the other hand, Content-Based methods make recommendation mainly based on users' and items' profile and contents [11]. The Content-Based methods can handle the item cold-start problem in recommendation because they directly extract information from the items' content and are independent of history information. Besides, Content-Based methods are often regarded as having a better explanation for recommendation, which is also very important in the recommender system.

2.2   Deep Learning in Recommendation

Traditional Content-Based methods are weak in processing complex text, image or audio information. Due to the remarkable performance deep learning gets in other fields recent years, many researchers are trying to introduce deep learning in recommender systems. Some are trying to strengthen the ability of Content-Based methods with deep learning. For example, Van used CNN to extract latent vectors from music audio and achieved excellent performance compared to principal components analysis (PCA) based method [12].

There are also some researchers who want to combine the deep learning with Collaborative Filtering. One way is to optimize the latent vector in matrix factorizing based methods. For example, Marginalized Denoising Auto-encoder based Collaborative Filtering (mDA-CF) [13] combines probabilistic matrix factorization with marginalized denoising stacked auto-encoders.

But these models still can not apply deep learning independently with traditional methods. Some works tried to do Collaborative Filtering alternatively by deep neural network [14–16]. The Neural Collaborative Filtering (NCF) model proposed by He et al. achieved significant improvements over many state-of-the-art methods. Besides, Google recently proposed a novel recommender model, Wide & Deep Learning, a state-of-the-art Content-Based method we would like to focus on in this paper.

3   Wide & Deep Learning

The Wide & Deep Learning was proposed by Google, whose motivation was to combine the advantages of deep neural network and linear model. Deep neural network has a strong ability to process sparse features like text and can extract dense embeddings which contain much information and are easier to use. It is considered to have a strong ability to generalize. But sometimes the deep model may over-generalize and recommend some less relevant items to the users, which is undesirable in the recommender system. One way to solve this problem is to combine the deep neural network with the linear model, which is considered to have a strong ability to memorize. It can learn a direct relationship between input features and output targets.

Google combines the deep neural network and linear model with logistic regression, shown in Fig. 1, where y_{ui} and $\hat{y}_{ui}$ denote the real and predicted label respectively.

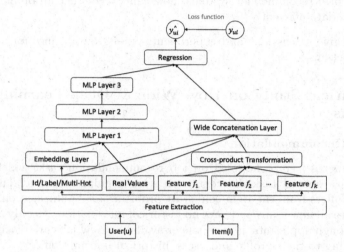

Fig. 1. The structure of Wide & Deep Learning

To formalize the prediction, let Y denote the class label and $\mathbf{x}$ denote the original features, then we have:

$$P(Y = 1|\mathbf{x}) = \sigma(\mathbf{w}_{wide}^\top [\mathbf{x}, \phi(\mathbf{x})] + \mathbf{w}_{deep}^\top a^{(l_f)} + b) \tag{1}$$

where $\sigma(\cdot)$ is the sigmoid function, $\phi(\mathbf{x})$ are the cross product transformations of the original features, and b is the bias. $\mathbf{w}_{wide}$ is the vector of all wide model weights, and $\mathbf{w}_{deep}$ is the weights applied on the final activations $a^{(l_f)}$.

The cross product transformations on the linear part are based on experience added artificially. It represents features' simultaneous appearance. For example, we have a 2-dimension one-hot vector for one's gender, and a 2-dimension one-hot vector for one's country (if Germany or not). Then we can generate a 4-dimension one-hot vector, which represents if one is a German man, woman or other countries' man or woman. We can define the cross-product transformation as follows:

$$\phi_k(\mathbf{x}) = \prod_{i=1}^{d} x_i^{c_{ki}} \quad c_{ki} \in \{0, 1\} \tag{2}$$

where ϕ_k represents the k-th transformation, c_{ki} is a boolean variable denoting whether ϕ_k is related to the i-th feature x_i.

We implemented the model with Tensorflow[1]. But there are still many questions on the model and how to take advantage of it:

1. Is feature selection necessary in the model and how to do it?
2. How do the deep neural network and linear model work with each other?
3. How can we avoid over-fitting of one part but at same time under-fitting of another part?
4. In real-world recommender systems, how can we conduct an online training with new data efficiently?

We will give discussions and experiments results on all questions above in following sections.

4   Empirical Study on How Wide & Deep Learning Works

4.1   Job Recommendation

Our experiments were conducted on job recommendation task, which can be regarded as a special task different from some regular recommendation task because it suffers from the item cold-start problem. Jobs published on websites every day are all new items and have no historical information. As only as a job is taken by enough applicants, it is no longer available. New jobs are recommended only according to their profile and users' historical information.

The dataset is from RecSys Challenge 2017[2]. The challenge is comprises into two parts, offline and online test. The size of training data and target data is shown in Table 1.

Table 1. The size of data in RecSys Challenge 2017

	Training			Target	
	Users	Items	Interactions	Users	Items
Offline test	1,497,020	1,306,054	322,766,002	74,840	46,559(41047[a])
Online test	981,673	1,037,282	92,949,362	48,167	1k–15k(all[a])

[a] Number of cold-start items

The goal is to recommend jobs to users. There is a specific algorithm defined by the organizers of the challenge to calculate the score. U denotes the target users and I denotes the target items. I_u denotes the items recommended to user $u \in U$ and U_i denotes the users item $i \in I$ is recommended to. Then we have:

$$Score = \sum_{u \in U} \sum_{i \in I_u} userSuccess(i, u) + \sum_{i \in I} itemSuccess(i) \tag{3}$$

---

[1] https://www.tensorflow.org.
[2] http://2017.recsyschallenge.com/.

in which

$$userSuccess(i, u) =$$
$$(1 * click(i, u) + 5 * reply\_or\_bookmark(i, u) + 20 * recruiter(i, u)$$
$$- 10 * delete\_only(i, u)) * (premium(u) + 1)$$
$$itemSuccess(i) = \tag{4}$$
$$(|\{userSuccess(i, u)|userSuccess(i, u) > 0, u \in U_i\}| > 0)$$
$$* 25 * (paid(i) + 1)$$

The size of users and items are too big for matrix factorization based methods. Collaborative Filtering methods do not work on the cold-start problem. Besides, the recommendation can also be considered as a classification problem, which all user-item pairs have a binary class label, like or dislike. As a result, Wide & Deep Learning, a state-of-the-art Content-Based method, is suitable.

4.2   Feature Selection for Wide & Deep

The deep neural networks are usually thought to have the ability to extract and select features automatically. For example, they can capture the lines and edges to understand the whole image. All we need to feed into them is the raw presentation of the images. However, Wide & Deep Learning is not a traditional end-to-end use of neural network to process image or text information but a non-linear classifier, whose input still includes features extracted by users of the model. We wonder if feature selection is still necessary.

Many item features and user features were extracted. Especially, though we don't have the history of some items, we can take advantage of user history. We extract many pairwise features, e.g.: which item tag or other attributes did user prefer in the history. Most of them are numeric features which usually are considered really important in the classifiers. Categorical features like country and discipline first pass through an embedding layer and then concatenate with other features as the input of hidden layers in deep part. Cross features we generate are mainly cross transformations between discipline and industry.

We compared the performance between the models with and without feature selection. To conduct feature selection, we calculated the Pearson correlation between the label and all the numeric features, and remove all the features whose Pearson correlation are lower than 0.2. Some experimental results on feature selection are shown in Table 2.

Parameters like learning rate, number and size of hidden layers are all well tuned separately in our models. We use Adam [17] as the optimizer and Cross-Entropy [18] as the loss function. Dropout has been tried but did not show remarkable effect. Early stop is conducted in the training process.

It can been seen that the performance after feature selection is much better than using all features, which show that feature selection is necessary to Wide & Deep Learning. Some features may be harmful to the network. Besides, historical pair features we extract are really important. They are necessary to solve the item

Table 2. Experiments on feature selection

	All features	No Historical[a]	Feature selection
Score	60	903	26,855

[a] The difference with "Feature Selection" is that there are no historical pair features which capture how the user prefers the item's tags or other attributes in the history.

cold-start problem for that they match the item attributes with users' history. Experiments in the following sections are all done with feature selection.

4.3   Roles of Wide and Deep Components in Learning

The two components of Wide & Deep are really different. They should play different roles in the model. Which part is more important and if they have different understandings to the input features are still unknown. Deep neural networks are always been regarded as black boxes, but there are new technologies can help understand them.

Layer-wise Relevance Propagation (LRP) is a state-of-the-art technique used in the field of image processing [19,20] and natural language processing [21,22]. It has been used to visualize what the neural network has learned and what it thinks is important. It back-propagates the LRP relevance from the higher layers to the lower layers according to the weight parameters to find out which part contributes most to the final result. Here we would like to introduce LRP to Wide & Deep to help us understand the network and better take advantage of it.

Feed-forward propagation in our Wide & Deep network can be defined as

$$x_j^{l+1} = \sigma(\sum_i x_i^l w_{ij}^{l,l+1} + b_j^{l+1}), \quad e.g. \quad \sigma(z) = relu(z) = max(0, z) \quad (5)$$

where j is the index of a particular neuron at layer $l + 1$, $w_{ij}^{l,l+1}$ and b_j^{l+1} are elements of weight matrix and bias from layer l to layer $l+1$. $\sigma(\cdot)$ is the activation function. Make R_i^l be the relevance score of i-th neuron at layer l, then we have

$$R_i^l = \sum_j \frac{z_{ij}}{\sum_k z_{kj} + \epsilon sign(\sum_k z_{kj})} R_j^{l+1} \quad with \quad z_{ij} = x_i^l w_{ij}^{l,l+1} \quad (6)$$

where ϵ is a small positive number to make the computation more stable and have better numerical properties.

We focus on two questions: Which part is more important? Does the wide part have better ability of memorizing sparse features' occurrence?

- The total relevance scores of deep part and wide part were first calculated, as shown in Table 3.

Table 3. Overall relevance score of deep part and wide part

	Deep	Wide
LRP Relevance	0.748	0.252

It can be seen that the deep neural network plays a much more important role in the Wide & Deep in this problem. It is reasonable because deep neural networks are supposed to have a better descriptive ability and are able to capture the relationship between features. To verify the importance of deep model, we also change the position of numeric features (which are the most important features in the network). Experiment results are shown in Table 4. A score is fed back for each upload. We show the average of 5 uploads.

Table 4. Experiments on numeric features' positions. ("Deep-Only"/"Both" is significantly better than "Wide-Only" with p-value $9e-5/3e-4$. But the difference between "Deep-Only" and "Both" is not significant in terms of p-value 0.15.)

	Wide-Only	Deep-Only	Both
Offline score	24,163	25,873	25,568

The result shows that if the deep part lacks important features, the performance of the entire model will significantly drop. If these features are fed into both wide and deep part, the performance is almost the same as not feeding into the wide part. This also verifies that deep part is the main part.

- Detailedly, the LRP relevance scores of features in two parts were calculated. The results are shown in Fig. 2.

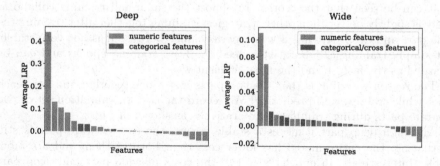

Fig. 2. LRP relevance of some features in two parts.

Although the feature with the highest relevance is a numeric feature in both parts, categorical and cross features have greater impacts in wide part than in the deep part. Since they are one-hot vectors which indicate the occurrence of properties, it shows that wide linear part memorizes the direct relationship between the occurrence and final label.

4.4   Comparative Study on Training Strategies

Different models have different learning speed on the same data. We want to investigate if it is a good way to combine the two kinds of model and train together. One of the concerns is that when one of the two parts have over-fitted, another part is still under-fitting. To avoid this, we tried different strategies of joint-training, including training deep part after training together or training together after training the two single parts. In each training period, we stop training when the cost on validation set doesn't get lower for 10 epochs. The average offline upload score and training epochs of 3 repeated experiments are shown in Table 5.

Table 5. Offline upload score of different joint-training strategy

Strategy[a]	Offline score	Epochs
Wide→Deep→Collaboratively	22,938	77
Deep→Wide→Collaboratively	25,354	50
Collaboratively→Deep	25,513	46
Collaboratively→Wide	25,727	34
Collaboratively	**25,786**	**22**

[a] When training a single part, we keep another part unchangeable and optimize the cost of the whole model. Otherwise, collaboratively training update the parameters of two parts at the same time.

It can be seen that the scores are almost the same and training collaboratively is slightly better than other strategies. Training together after training the wide part and deep part has a worse performance because it is not very stable. But simply training collaboratively need the least training epochs and can be regarded as the fastest and the most efficient way.

The reason we think is that the two parts may work together and help each other while training and predicting. We record the cost on validation set of two separate parts during training collaboratively, as shown in Fig. 3.

Although deep part achieves a stable status, the cost of wide part is still fluctuating, the entire model has lower cost than both of them and are more stable. Interestingly, in epochs 8 and 9, the costs of wide part and deep part fluctuate at different directions, but the cost of entire model drops. The Wide & Deep Learning is different from ensemble learning. Joint-training in Wide & Deep makes the two parts help and learn from each other during training collaboratively.

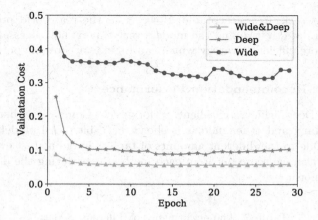

Fig. 3. Cost on validation set of two parts (Wide/Deep) while training.

4.5   Online Training

In real-world recommender system, new data is coming every day. Preferences of users and attributes of items may change over time and influence each other [23]. How to conduct an online training and update the model regularly is important so that it can keep a good performance.

In this job recommendation task, data provided in the online test is different from that provided in the offline test. The new data contains two kinds of information: new training data and everyday interaction feedback. The most straightforward way is to re-train the recommendation model every day. However, it may cost too much time and computing resources. Another way is to update the model in an incremental way, which means we can load the previous model and continue training with fresh data. It is necessary to choose an online training strategy in the challenge. Since that only one strategy can be upload each day and the number of target items differs between days, it is unfair to compare the scores calculated from feedbacks whose amount is also changeable. To conduct a fair comparison, we generate results on one day's targets with enough feedback and measure the performances on our own. The precision (positive interaction number/impression number) of different model update strategies is calculated. The results are shown in Table 6.

Table 6. Performance of different model update strategies

Model	Online-only	Fine-tuning[a]	Re-train
Precision	0.174	0.311	0.384

[a] Load original model with best performance in the offline test and train with new data.

It makes sense that the re-trained model gets the best performance, but actually, the time of training is almost two times longer than the fine-tuning model. In this task, it is acceptable. However in the real-world problem with larger scale of data, re-training the model every time is not necessary and fine-tuning is a more efficient strategy which can provide acceptable performance.

4.6 Overall Recommendation Performance

The Wide & Deep achieves excellent performance. Comparison between Wide & Deep Learning and other models is shown in Table 7. All models are tuned at our best. Due to the different amounts of targets in online test every day, we calculate the average score per item each method gets during the days to show an overall performance.

Table 7. The performance of different models

	Offline score	Online score/Item
Item-neighbour	12,438	1.0180
Historical Enhancement	20,450	0.9541
XGBoost	14,628	—
Logistic Regression (Wide)	24,168	—
Neural Network (Deep)	25,539	—
Wide & Deep	**26,855**	**1.0210**

The performance of Wide & Deep Learning is better than other models. Especially, the Wide & Deep achieves better performance than both the wide model and the deep model, which shows that the combination is reasonable.

5 Conclusion

In this paper, our aim is not to propose a new model, but to conduct an empirical insight study of a novel model, Wide & Deep Learning, in real scenario with large scale online data. We want to better understand and take advantage of it. Although the experiments were done in only one field, job recommendation, the datasets in offline and online test are different. The offline dataset includes some instances artificially added by organizers and the online test has a more real evaluation. The two tests can be regarded as two separated problems. Our contributions can be concluded as follows:

1. We found that feature selection is still necessary in this model. Irrelevant features are harmful to the model's performance.
2. We used LRP to analyze the network and results shows that deep part takes an important role in the network, important features like numeric features must not be absent in the deep neural network.

3. We conclude that joint-training the wide and deep parts collaboratively is the most efficient and effective way.
4. We discussed the strategies to conduct an online training. Re-training the model with whole data set is best with enough time and computing resources. Loading the old model and continue to train is more efficient cost of a little lower performance.

As a state-of-the-art Content-Based method, Wide & Deep combines the generalizing ability of deep neural network and memorizing ability of the wide linear model. But the structure currently lacks the ability to process text or image information. How to import CNN or other kinds of the network to the model and apply it to other datasets and problems is a good question and will be included in our future work.

References

1. Sweeney, L.: Confidentiality, disclosure, and data access: theory and practical applications for statistical agencies. Inform. Explosion, 43–74 (2001)
2. LeCun, Y., Bengio, Y., Hinton, G.: Deep learning. Nature **521**(7553), 436–444 (2015)
3. He, K., Zhang, X., Ren, S., et al.: Deep residual learning for image recognition. In: Proceedings of the IEEE Conference on Computer Vision and Pattern Recognition, pp. 770–778 (2016)
4. Manning, C., Socher, R., Fang, G.G., et al.: CS224n: natural language processing with deep learning1 (2017)
5. Son, L.H.: Dealing with the new user cold-start problem in recommender systems: a comparative review. Inform. Syst. **58**, 87–104 (2016)
6. Cheng, H.T., Koc, L., Harmsen, J., et al.: Wide & deep learning for recommender systems. In: Proceedings of the 1st Workshop on Deep Learning for Recommender Systems, pp. 7–10. ACM (2016)
7. Pan, R., Zhou, Y., Cao, B., et al.: One-class collaborative filtering. In: Eighth IEEE International Conference on Data Mining, ICDM 2008, pp. 502–511. IEEE (2008)
8. Liang, D., Charlin, L., McInerney, J., et al.: Modeling user exposure in recommendation. In: Proceedings of the 25th International Conference on World Wide Web. International World Wide Web Conferences Steering Committee, pp. 951–961 (2016)
9. Weimer, M., Karatzoglou, A., Smola, A.: Improving maximum margin matrix factorization. Mach. Learn. **72**(3), 263–276 (2008)
10. Rendle, S., Freudenthaler, C., Gantner, Z., et al.: BPR: Bayesian personalized ranking from implicit feedback. In: Proceedings of the Twenty-Fifth Conference on Uncertainty in Artificial Intelligence, pp. 452–461. AUAI Press (2009)
11. Lops, P., De Gemmis, M., Semeraro, G.: Content-based recommender systems: state of the art and trends. In: Ricci, F., Rokach, L., Shapira, B., Kantor, P. (eds.) Recommender Systems Handbook. Springer, Boston (2011). doi:10.1007/978-0-387-85820-3_3
12. Van den Oord, A., Dieleman, S., Schrauwen, B.: Deep content-based music recommendation. In: Advances in Neural Information Processing Systems, pp. 2643–2651 (2013)

13. Li, S., Kawale, J., Fu, Y.: Deep collaborative filtering via marginalized denoising auto-encoder. In: Proceedings of the 24th ACM International on Conference on Information and Knowledge Management [S.l.], pp. 811–820. ACM (2015)
14. He, X., Liao, L., Zhang, H., et al.: Neural collaborative filtering. In: Proceedings of the 26th International Conference on World Wide Web. International World Wide Web Conferences Steering Committee, pp. 173–182 (2017)
15. Zheng, Y., Tang, B., Ding, W., et al.: A neural autoregressive approach to collaborative filtering. arXiv preprint arXiv:1605.09477 (2016)
16. Zheng, Y., Liu, C., Tang, B., et al.: Neural autoregressive Collaborative Filtering for implicit feedback. In: Proceedings of the 1st Workshop on Deep Learning for Recommender Systems [S.l.], pp. 2–6. ACM, 2016
17. Kingma, D., Adam, B.J.: A method for stochastic optimization. arXiv preprint arXiv:1412.6980 (2014)
18. Deng, L.Y.: The cross-entropy method: a unified approach to combinatorial optimization. Monte-Carlo Simul. Mach. Learn. (2006)
19. Bach, S., Binder, A., Montavon, G., et al.: On pixel-wise explanations for non-linear classifier decisions by layer-wise relevance propagation. PloS One **10**(7), e0130140 (2015)
20. Binder, A., Bach, S., Montavon, G., Müller, K.-R., Samek, W.: Layer-wise relevance propagation for deep neural network architectures. Information Science and Applications (ICISA) 2016. LNEE, vol. 376, pp. 913–922. Springer, Singapore (2016). doi:10.1007/978-981-10-0557-2_87
21. Ding, Y., Liu, Y., Luan, H., et al.: Visualizing and understanding neural machine translation
22. Arras, L., Montavon, G., Müller, K.R., et al.: Explaining recurrent neural network predictions in sentiment analysis. arXiv preprint arXiv:1706.07206 (2017)
23. Dai, H., Wang, Y., Trivedi, R., et al.: Recurrent coevolutionary feature embedding processes for recommendation. arXiv preprint arXiv:1609.03675 (2016)

A Neural Network Model for Social-Aware Recommendation

Lin Xiao[1]([✉]), Zhang Min[2], Liu Yiqun[2], and Ma Shaoping[2]

[1] Institute of Interdisciplinary Information Sciences,
Tsinghua University, Beijing, China
jackielinxiao@gmail.com
[2] Tsinghua National Laboratory for Information Science and Technology,
Department of Computer Science and Technology, Tsinghua University,
Beijing 100084, China
{z-m,yiqunliu,msp}@tsinghua.edu.cn

Abstract. Social-aware recommender systems have been popular with the rapid growth of social media applications. Existing approaches have attempted to accommodate social information into typical Collaborative Filtering methods and achieved significant improvements. Neural networks are gaining increasing interests in information retrieval tasks. However few studies have considered applying neural network in social-aware recommendation tasks. In this paper, we aim to fill this gap and propose a social-aware neural recommender system. Extensive experiments on real-world datasets demonstrate that our model outperforms state-of-art approaches significantly.

1 Introduction

Recommender systems aim to provide information or items of interest to users. The various user profiles have been of great importance in enhancing the performance of recommendation. The rapid growth of social media applications reflect the social connections between users, which contributes to a better understanding of user preferences. Existing studies have attempted to utilize social information in recommender systems and achieved significant improvement.

The classical social network theory builds upon two important assumptions: first, users who are socially connected are believed to be more similar than those who are not. This is also referred to as "Homophily Effect". Second, users are connected with different social ties. These social theories have been exploited in recommender systems in previous studies.

Meanwhile, the neural networks are widely applied in various fields, including computer vision, natural language processing and information retrieval. The strong expressive power of neural networks allows for extracting useful features from the input and further adapts them into the corresponding tasks. The studies on applying neural networks in recommender systems can be found in [15]. The typical approaches in previous studies are built upon Restricted Bolzman Machine and its variants. Recent studies attempt to mimic collaborative filtering

W.-K. Sung et al. (Eds.): AIRS 2017, LNCS 10648, pp. 125–137, 2017.
https://doi.org/10.1007/978-3-319-70145-5_10

and matrix factorization with neural networks. Each user and item are embedded as corresponding vectors and factorization is conducted on these vectors following classical approaches.

There exist some studies on network embedding [19,28], including social networks and information networks. Similar to word embedding [9] in NLP, the network embedding aims to represent each node as a vector in the Euclidean space and the nodes connected with edges are closer to each other. The network embedding problem is close to recommendation when each user and item are seen as nodes and the interactions between them are seen as edges. However, it is not trivial to apply network embedding techniques on recommendation, especially in social-aware recommendation tasks. The interactions between users and items are different from those between users, therefore the network embedding procedure can not be directly adapted into two heterogeneous networks directly. Meanwhile, social information is referred to as the contents of the node while network embedding is to discover the structures of the network. Therefore social-aware recommendation tasks require a modeling of the network structure with node contents.

Despite existing successes in applying neural networks on recommender systems and network embedding, no previous studies considered the specific problem of incorporating social information into the recommender systems with neural networks. There are two important challenges in designing social-aware neural recommender systems: first, it is unknown how to encode social information into the neural network framework; second, the impact of social connections needs to be modeled properly in the recommendation process. Although existing studies attempt to utilize social network theories in recommendation, the theories have not been well accommodated into neural network frameworks.

In this work, we aim to extend the classical SVD++ model with both social information and the powerful neural network framework. The classic SVD++ model is proposed in [6] and gets widely used in recommendation for its superior performances in Netflix Challenge and real-life recommender systems. The basic idea of SVD++ is to model both explicit and implicit feedbacks with matrix factorization. We treat social information as the implicit feedbacks in SVD++ and further model its interaction with users and items with factorization and neural network scheme. Extensive experiments have been conducted on two real-world datasets, the results validate the soundness of coordinating social information with neural networks.

The remainder of this paper is organized as follows: Sect. 2 introduces some important aspects of related studies; the models of SVD++ and Neural SVD++ are introduced in Sect. 3; the experimental results are presented in Sects. 4 and 5 presents the conclusion of this paper and provides a discussion about the future work.

2 Related Work

In this section, we make a brief review of related studies, including the typical Collaborative Filtering approach, the social-aware recommender systems and existing studies on recommendation with neural networks.

2.1   Collaborative Filtering for Recommendation

Collaborative Filtering (CF) is a typical approach for recommendation [16]. The motivation comes from the assumption that people often get the best recommendations from someone with tastes similar to themselves. There are two generic Collaborative Filtering approaches, i.e. user based CF and item based CF. The user-based CF adopts the motivation stated before while the item-based CF assumes that items tend to receive similar ratings with other similar items.

Existing CF approaches include two categories: memory-based and model-based approaches. User-KNN and Item-KNN are two representative algorithms in memory-based algorithms. Locality-sensitive hashing [17] is a typical algorithm adopted to find similar users, which implements the KNN algorithm in linear time.

Model-based approaches use machine learning techniques to model the generation process of ratings. Among various model-based CF methods, Matrix Factorization (MF) is the most popular and effective one, which assumes that users and items are represented as vectors in a latent factor space. Some MF based approaches, including SVD++ [6], NMF (Non-Negative Matrix Factorization) [26], MMMF (Max-Marginal Matrix Factorization) [13], BMF (Biased Matrix Factorization) [6] and PMF (Probabilistic Matrix Factorization) [14] have achieved superior accuracy and scalability in recommendation due to the dimension reduction nature.

2.2   Social-Aware Recommendation

Social information is known to be helpful in recommendation systems [3,4,10, 11,20,25]. Most studies assume that some social homophily effect exists, causing users to behave consistently with others in social connections [3,10].

[5] extends the approach in [4] by combining random walks with collaborative filtering for item recommendation. The Multi-Relational Bayesian Personalized Ranking (MR-BPR) model [7], which combines multi-relational matrix factorization with the BPR framework, predicts both user feedback on items and on social relationships.

There are two state-of-the-art algorithms that utilize social information for item recommendation tasks [27]. In [27], the authors assume that users are more likely to have seen items consumed by their friends, and use this effect to sample negative feedback in BPR. In contrast, [1] utilizes Poisson factorization to incorporate social information into a matrix factorization scheme. None of the aforementioned studies considers the modeling of missing feedback in social relationships. In the present study, we use some of the most popular social-aware recommendation algorithms as benchmarks.

A recent study [21] considers the strength of social ties and its application in social recommendations. The neighborhood overlap is used to approximate tie strength and extend the popular BPR model to incorporate the distinction between strong and weak ties. As this is an extension of BPR, the missing feedback is randomly selected as negative feedback. In [12], the Collaborative

Ranking (CofiRank) model [22] is extended to include social connections. As CofiRank considers observations of both positive and negative feedback, the issue of missing feedback is not investigated. Two different CofiRank strategies are proposed based on the notions of Social Reverse Height and Social Height, which quantify how well the relevant and irrelevant items of users and their social friends have been ranked, respectively.

2.3 Neural Network for Recommendation

Another trend on related research is to utilize deep learning method in recommender system. With the growing research in artificial neural networks and deep learning techniques, some of the famous Deep Learning models are applied in recommender systems. The first model [15] applied in recommender system is the Restricted Botzman Machine which assumes each user is depicted with a RBM where the ratings are modeled as binary input vectors and the hidden units and correlation weights are used to generate the full ratings. Another dual-reversible RBM which takes linear inputs is proposed in [2] and the model uses a single RBM instead of user-size RBMs to generate the full ratings. Despite that the RBM technique can achieve performances that are comparable to popular Matrix Factorization techniques, the training process of RBM is quite intractable. The mean-field technique is used to relax the model into a two-layer feed-forward neural network and thus the model is quite easy to train. The RBM based model is referred as Neural Auroregressive Distribution Estimator (NADE) [29]. The model is applied in modeling the distribution of high-dimensional vectors in [8] and achieves better performance than the original RBM model. Recently, a new model is proposed in [18] which is a novel autoencoder framework for collaborative filtering (CF) since auto-encoder has a good performance on dimension reduction. However existing ANN (Artificial Neural Network) based recommender systems have not started to combine side information with recommender systems and a proper model is yet to be designed in the future.

3 Model

In this section, we specifically introduce the Neural Social SVD++ (Neural SSVD++) model. Before we present the details of the model, some preliminary knowledge about SVD++ is first introduced. Then we introduce the Neural SVD++ model and how we extend the idea of SVD++ with the Neural Network framework.

3.1 SVD++

SVD++ is first proposed in [6]. In this model, each user and item and implicit feedback are represented as latent factor vectors and the rating is assumed to be a combination of user and item latent factor vectors with implicit feedbacks. The implicit feedbacks refer to the interaction records of users, i.e. the clicks of advertisements by users, the browsing history of users on the websites. In SVD++, the influence of these implicit feedbacks are considered and modeled as latent factor vectors in the matrix factorization framework. The parameters of this model is listed in Table 1.

$$\hat{r}_{ui} = b_{ui} + q_i^T (p_u + |N(u)|^{-\frac{1}{2}} \sum_{j \in N(u)} y_j) \tag{1}$$

where $b_{ui} = b_u + b_i + \mu$. The parameters of this model can be determined by minimizing the empirical squared loss functions of observed ratings, which can be achieved with Stochastic Gradient Descent algorithm.

$$\min. \sum_{u,i \in \mathcal{O}} |\hat{r}_{ui} - r_{ui}|^2 + \lambda(\sum_i \|q_i\|_2 + \sum_u \|p_u\|_2 + \sum_j \|y_j\|_2) \tag{2}$$

Table 1. Variables of SVD++

Variables	Meaning
r_{ui}	The rating given by user u to item i
$\hat{r}_{ui}$	The prediction of rating r_{ui}
$\mathcal{O}$	The set of observed ratings
b_u	The bias of user u
b_i	The bias of item i
μ	The global bias
q_i	The latent factor vector of item i
y_j	The latent factor vector of implicit feedback j
p_u	The latent factor vector of user u
$N(u)$	The set of implicit feedbacks of user u

3.2 TrustSVD

TrustSVD model [3] is also built on top of the SVD++ model, which also takes into consideration user/item biases and the influence of rated items other than user/item-specific vectors on rating prediction. Formally, the rating for user u on item i is predicted by:

$$\hat{r}_{ui} = b_u + b_i + \mu + q_i^T (p_u + |I_u|^{-\frac{1}{2}} \sum_{j \in I_u} w_j + |T_u|^{-\frac{1}{2}} \sum_{v \in T_u} y_u) \tag{3}$$

where $\{y_v, \forall v \in T_u\}$ refer to user-specific latent factor vector of users $v \in T_u$ who are socially connected to user u, and w_j denotes the influence of items $j \in I_u$ rated by user u in the past.

This model is a direct extension of SVD++ with trust relationships. The parameters can be learnt via SGD in a similar manner.

3.3   Neural SSVD++

We extend the idea of SVD++ with Neural Network and propose the model Neural Social SVD++ (Neural SSVD++). The model is shown in Fig. 1 (Those arrows in dash lines refer to the element-wise product of vectors). Following a similar spirit of SVD++, Neural SSVD++ embeds users, items and socially connected users (seen as implicit feedbacks in SVD++) as latent factor vectors. The impact of social connections is incorporated into the model by considering both its representation in the latent factor space and its interaction with users. Moreover, we utilize the nonlinearity of neural networks to model the relationship between ratings and the latent factor vectors and their interactions.

In this model, the input is a row vector containing three components: the first component is the target user ID u and the second component is the target item ID i, the third component is a sequence of user ID $N(u)$ where each user is socially connected to the target user. The input then goes through embedding layer and each target user and item are mapped into corresponding vectors, the sequence of socially connected users are mapped into a sequence of vectors $y_j, \forall j \in N(u)$. Then the sequence of social vectors is reduced to a single vector $|N(u)|^{-1} \sum_{j \in N(u)} y_j$. The interactions between these latent factor vectors are modeled in a pairwise manner: $p_u \otimes q_i$ and $|N(u)|^{-1} \sum_{j \in N(u)} y_j \otimes q_i$ (where $\otimes$ is an element-wise product operator). Then we concatenate these factor vectors together: $H_{ui} = [p_u; q_i; |N(u)|^{-1} \sum_{j \in N(u)} y_j; p_u \otimes q_i; |N(u)|^{-1} \sum_{j \in N(u)} y_j \otimes q_i]$. After the concatenation, we use a two-layer fully connected layer to model the

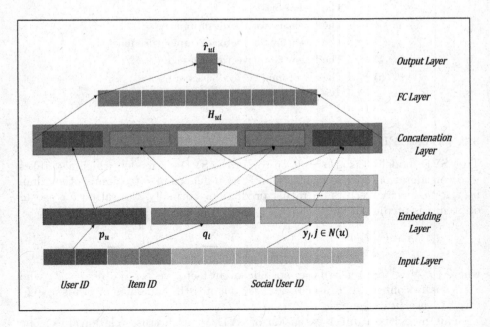

Fig. 1. Neural Social SVD++, all the activation functions are Relu.

relationship between the concatenated vector and the output rating with non-linearity:

$$y_{ui} = Relu(W_2 \cdot Relu(W_1 \cdot H_{ui} + b_1) + b_2) \tag{4}$$

Since H_{ui} is a concatenation of multiple latent factor vectors, the function can be re-written into following form:

$$y_{ui} = Relu(W_2 \cdot Relu(W_{11}p_u + W_{12}q_i + W_{13}|N(u)|^{-1} \sum_{j \in N(u)} y_j +$$

$$W_{14}p_u \otimes q_i + W_{15}|N(u)|^{-1} \sum_{j \in N(u)} y_j \otimes q_i + b_1) + b_2) \tag{5}$$

For the two Fully Connected layers, we first reduce the dimension of hidden units to 16 and then reduce it to the single valued rating $\hat{r}_{ui}$.

3.4 Learning

We adopt the typical back-propagation method for learning the parameters of the model. For the rating prediction tasks, we use the empirical squared loss function as the objective and use Adagrad algorithm for optimization:

$$\mathbb{L} = \sum_{u,i \in \mathcal{O}} |\hat{r}_{ui} - r_{ui}|^2 + \lambda \Omega(W, p, q, y) \tag{6}$$

where the first component is the squared loss of rating prediction and the second component is the regularization term.

4 Experiment

In this section, we present the experimental results on two real-world datasets and the performances in different circumstances. Moreover, we compare the performances of Neural SSVD++ with and without social information as input. The results show that our model outperforms other baselines and social information contributes to the prediction accuracy significantly.

4.1 Experiment Setting

Two representative datasets are selected for experiments: Ciao and Epinions. These two datasets contain both rating records and in-site social relationships The statistics of these two datasets are presented in Table 2.

We adopt five-fold cross-validation, the datasets are split into five folds and four of them are selected as training set while the remaining fold is used for testing. The experiment is conducted in an environment with a 1.8 GHZ CPU. We use the mini-batch training scheme when training the model and the batch size is set to 128.

We select some state-of-the-art approaches as baseline algorithms, including SoRec, Soreg, TrustSVD, TrustMF and SVD++:

Table 2. Statistical details of the datasets

Datasets	#Users	#Items	#Ratings	#Links	Link type
Ciao	7,267	11,211	147,995	111,781	Unilateral
Epinions	38,089	23,585	488,917	433,416	Unilateral

- **SoRec**[10]: Sorec co-factorizes the rating matrix and social matrix simultaneously and both matrices share the same user factor vectors.
- **SoReg**[11]: The model adds social regularization into the matrix factorization framework based on the social homophily effect.
- **SVD++**[6]: SVD++ is a model that merges latent factor model and neighbourhood effect together. Furthermore, it can be extended to incorporate both implicit and explicit feedbacks from users.
- **TrustMF**[24]: TrustMF assigns each user a trustor-specific vector and a trustee-specific vector. The model can choose to incorporate either vector or both vectors in the matrix factorization framework.
- **TrustSVD**[3]: TrustSVD extend SVD++ with social trust information and takes into account both the explicit and implicit influence of ratings and trust information when predicting ratings of unknown item.

In order to evaluate the performances on rating prediction tasks, we use typical MAE and RMSE metrics:

$$MAE = \frac{1}{|T|} \sum_{(u,i) \in T} |\hat{r}_{ui} - r_{ui}|$$

$$RMSE = \sqrt{\frac{1}{|T|} \sum_{(u,i) \in T} |\hat{r}_{ui} - r_{ui}|^2}$$

$$(7)$$

where T denotes the testing set, $\hat{R}_{ij}$ denotes the prediction of the ground truth R_{ij}.

4.2 Experimental Results

We list the comparison between Neural SSVD++ and other baselines under two circumstances in Tables 3 and 4. First, we compare the performances of Neural SSVD++ with other baselines on two datasets. The results indicate that Neural SSVD++ outperforms other baselines in terms of rating prediction accuracy. Generally, the social-aware recommendation algorithms perform better than those social-unaware algorithms (comparing SVD++ with others), which indicates that social connections are informative for recommendation. More specifically, Neural SSVD++ outperforms other social aware algorithms, which means that accommodating social information with neural networks significantly improves the expressive and predictive power of the model.

Table 3. Performance of Rating Prediction on Ciao and Epinions, *: $p < 0.01$

Ciao	SoRec	SoReg	TrustSVD	TrustMF	SVD++	Neural SSVD++
MAE	0.761	0.815	0.723	0.742	0.748	**0.690**∗
MAE Improvement	9.33%	15.34%	4.56%	7.01%	7.75%	-
RMSE	1.010	1.076	0.956	0.983	1.001	**0.942**∗
RMSE Improvement	6.73%	12.45%	1.46%	4.17%	5.89%	-
Epinions	SoRec	SoReg	TrustSVD	TrustMF	SVD++	Neural SSVD++
MAE	0.882	0.932	0.805	0.818	0.818	**0.790**∗
MAE Improvement	10.43%	15.23%	1.86%	3.42%	3.42%	-
RMSE	1.114	1.232	1.044	1.095	1.057	**1.025**∗
RMSE Improvement	7.99%	16.80%	1.82%	6.39%	3.02%	-

Table 4. Performance on cold-start users in Ciao and Epinions, *: $p < 0.01$

Ciao	SoRec	SoReg	TrustSVD	TrustMF	SVD++	Neural SSVD++
MAE	0.730	0.949	0.721	0.752	0.749	**0.704**∗
MAE Improvement	3.56%	25.81%	2.36%	6.38%	6.00%	-
RMSE	0.998	1.214	0.962	1.096	1.020	**0.950**∗
RMSE Improvement	5.05%	21.75%	1.25%	13.32%	6.86%	-
Epinions	SoRec	SoReg	TrustSVD	TrustMF	SVD++	Neural SSVD++
MAE	0.846	1.139	0.868	0.853	0.889	**0.836**∗
MAE Improvement	1.18%	26.60%	3.69%	1.99%	5.96%	-
RMSE	1.138	1.437	1.105	1.125	1.162	**1.076**∗
RMSE Improvement	5.45%	25.12%	2.62%	4.36%	7.40%	-

Moreover, we evaluate the performances of Neural SSVD++ and other baselines w.r.t cold-start users. We adopt the setting where users rate fewer than five items are referred as cold-start users, which is typical in related studies like [3,23]. The results show that Neural SSVD++ still performs better than others. Since the utilization of social information compensates for the shortage of ratings from cold users, neural networks have an advantage of expressive power and work more effectively for cold-start users.

The Impact of Network Depth. Typically, the depth of neural network is related to the performance. In tasks like image classification and text classification, neural networks are designed to be deep for a better performance. For this task, we conduct experiments to explore the relationship between network depth and the rating prediction accuracy. Surprisingly, we discover that deeper network does not lead to superior performances necessarily. We alter the depth of Fully Connected layers from one to five and observe the resulting RMSE in different cases.

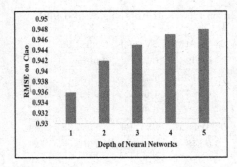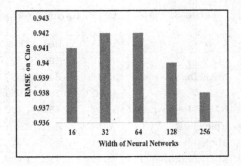

Fig. 2. RMSE of Neural Social SVD++ with different Depths and Widths

The RMSE of Neural SSVD++ under different network depths on Ciao are presented in Fig. 2. We discover that the performances of Neural SSVD++ does not necessarily get improved with deeper network. Since the depth of the network represents the model complexity, higher depth may cause overfitting in some cases.

The Impact of Embedding Dimensions. The dimension of embedded vectors decides the width of the neural network. We also evaluate the impact of embedding dimensions in Neural SSVD++. We alter the embedding dimension from 16 to 256 with a power of two for each trial. The MAE in different cases are shown in Fig. 3. Judging from the results, a proper setting for rating prediction tasks is a dimension of 64 or 128.

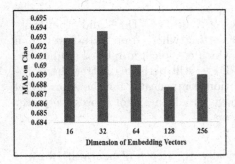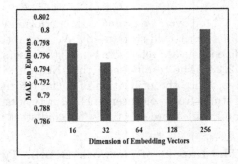

Fig. 3. MAE of Neural Social SVD++ with different Embedding Dimensions

Similarly, the performance does not keep going better with the increase of embedding dimensions. A greater dimension of embedding vectors allows for stronger expressive power, however also leads to higher risk of overfitting.

The Impact of Social Coordination. In order to evaluate the enhancement brought by social information, we also conduct experiments on Ciao dataset with Neural SSVD but removing the social input from the model. Therefore the predicted rating becomes:

$$y_{ui} = Relu(W_2 \cdot Relu(W_{11}p_u + W_{12}q_i + W_{14}p_u \otimes q_i + b_1) + b_2) \qquad (8)$$

We present the comparison between the performances of Neural SSVD++ with and without social input in Table 5. The results show that incorporating social information into the model leads to significant improvement. This illustrates the importance of introducing social context into recommendation. Despite that neural network framework preserves strong nonlinearity and expressive power, social information still makes a significant contribution to the improvement of recommendation accuracy.

Table 5. Performance of Neural SSVD++ on Ciao with/without Social Input, *: $p < 0.01$

Ciao	TrustSVD	SVD++	Neural SSVD++ without social	Neural SSVD++
MAE	0.723	0.748	0.729	**0.690***
RMSE	0.956	1.001	0.995	**0.942***

5 Conclusion

In this paper, we aim to incorporate social information into the recommender systems with the framework of Neural Networks. We extend the classical SVD++ model by introducing socially connected users as implicit feedbacks and extend the matrix factorization operations with neural network schemes. The strong expressive power of neural networks contributed to the modeling of the relationship between ratings and the latent factors. In the proposed Neural SSVD++ model, the advantages of SVD++ and neural networks are kept and combined in a same model simultaneously. Extensive experiments are conducted on two real-world datasets, Neural SSVD++ has achieved significant improvements over state-of-the-art social-aware approaches.

In the future work, we aim to design a social-specific neural network structure for social-aware recommendation. Since the social network forms a user-user graph and implicit structural knowledge may not be fully captured in current model. Another interesting direction is to exploit more information in social media applications and accommodate them into the model. This also allows for a broader use of neural network approaches in recommender systems when texts and images are available.

Acknowledgement. This work was supported by Natural Science Foundation (61532011, 61672311) of China and National Key Basic Research Program (2015CB358700).

References

1. Chaney, A.J., Blei, D.M., Eliassi-Rad, T.: A probabilistic model for using social networks in personalized item recommendation. In: Recsys (2015)
2. Georgiev, K., Nakov, P.: A non-iid framework for collaborative filtering with restricted Boltzmann machines. In: Proceedings of the 30th International Conference on Machine Learning (ICML-2013), pp. 1148–1156 (2013)
3. Guo, G., Zhang, J., Yorke-Smith, N.: TrustSVD: collaborative filtering with both the explicit and implicit influence of user trust and of item ratings. In: AAAI (2015)
4. Jamali, M., Ester, M.: Trustwalker: a random walk model for combining trust-based and item-based recommendation. In: KDD (2009)
5. Jamali, M., Ester, M.: Using a trust network to improve top-n recommendation. In: Recsys (2009)
6. Koren, Y.: Factorization meets the neighborhood: a multifaceted collaborative filtering model. In: Proceedings of the 14th ACM SIGKDD International Conference on Knowledge Discovery and Data Mining, KDD 2008, pp. 426–434. ACM, New York (2008). http://doi.acm.org/10.1145/1401890.1401944
7. Krohn-Grimberghe, A., Drumond, L., Freudenthaler, C., Schmidt-Thieme, L.: Multi-relational matrix factorization using Bayesian personalized ranking for social network data. In: WSDM (2012)
8. Larochelle, H., Murray, I.: The neural autoregressive distribution estimator. In: AISTATS, vol. 6, p. 622 (2011)
9. Levy, O., Goldberg, Y.: Neural word embedding as implicit matrix factorization. In: Advances in Neural Information Processing Systems, pp. 2177–2185 (2014)
10. Ma, H., Yang, H., Lyu, M.R., King, I.: Sorec: social recommendation using probabilistic matrix factorization. In: CIKM (2008)
11. Ma, H., Zhou, D., Liu, C., Lyu, M.R., King, I.: Recommender systems with social regularization. In: WSDM (2011)
12. Rafailidis, D., Crestani, F.: Joint collaborative ranking with social relationships in top-n recommendation. In: CIKM (2016)
13. Rennie, J.D., Srebro, N.: Fast maximum margin matrix factorization for collaborative prediction. In: Proceedings of the 22nd International Conference on Machine Learning, pp. 713–719. ACM (2005)
14. Salakhutdinov, R., Mnih, A.: Probabilistic matrix factorization. In: Advances in Neural Information Processing Systems 20, Proceedings of the Twenty-First Annual Conference on Neural Information Processing Systems, Vancouver, British Columbia, Canada, 3–6 December 2007, pp. 1257–1264 (2007). http://papers.nips.cc/paper/3208-probabilistic-matrix-factorization
15. Salakhutdinov, R., Mnih, A., Hinton, G.: Restricted Boltzmann machines for collaborative filtering. In: Proceedings of the 24th International Conference on Machine Learning, ICML 2007, pp. 791–798. ACM, New York (2007). http://doi.acm.org/10.1145/1273496.1273596
16. Sarwar, B., Karypis, G., Konstan, J., Riedl, J.: Item-based collaborative filtering recommendation algorithms. In: Proceedings of the 10th International Conference on World Wide Web, WWW 2001, pp. 285–295. ACM, New York (2001). http://doi.acm.org/10.1145/371920.372071
17. Satuluri, V., Parthasarathy, S.: Bayesian locality sensitive hashing for fast similarity search. Proc. VLDB Endowment 5(5), 430–441 (2012)

18. Sedhain, S., Menon, A.K., Sanner, S., Xie, L.: Autorec: autoencoders meet collaborative filtering. In: Proceedings of the 24th International Conference on World Wide Web, WWW 2015 Companion, pp. 111–112. ACM, New York (2015). http://doi.acm.org/10.1145/2740908.2742726

19. Tang, J., Qu, M., Wang, M., Zhang, M., Yan, J., Mei, Q.: Line: Large-scale information network embedding. In: Proceedings of the 24th International Conference on World Wide Web, pp. 1067–1077. International World Wide Web Conferences Steering Committee (2015)

20. Tang, J., Wang, S., Hu, X., Yin, D., Bi, Y., Chang, Y., Liu, H.: Recommendation with social dimensions. In: AAAI (2016)

21. Wang, X., Lu, W., Ester, M., Wang, C., Chen, C.: Social recommendation with strong and weak ties. In: CIKM (2016)

22. Weimer, M., Karatzoglou, A., Le, Q.V., Smola, A.J.: COFI RANK - maximum margin matrix factorization for collaborative ranking. In: NIPS (2008)

23. Yang, B., Lei, Y., Liu, D., Liu, J.: Social collaborative filtering by trust. In: IJCAI 2013, Proceedings of the 23rd International Joint Conference on Artificial Intelligence, Beijing, China, 3–9 August 2013 (2013). http://www.aaai.org/ocs/index.php/IJCAI/IJCAI13/paper/view/6750

24. Yang, B., Lei, Y., Liu, J., Li, W.: Social collaborative filtering by trust. IEEE Trans. Pattern Anal. Mach. Intell. 39(8), 1633–1647 (2016)

25. Ye, M., Liu, X., Lee, W.C.: Exploring social influence for recommendation: a generative model approach. In: SIGIR (2012)

26. Zhang, S., Wang, W., Ford, J., Makedon, F.: Learning from incomplete ratings using non-negative matrix factorization. In: SDM, vol. 6, pp. 548–552. SIAM (2006)

27. Zhao, T., McAuley, J., King, I.: Leveraging social connections to improve personalized ranking for collaborative filtering. In: CIKM (2014)

28. Zhao, W.X., Huang, J., Wen, J.-R.: Learning distributed representations for recommender systems with a network embedding approach. In: Ma, S., Wen, J.-R., Liu, Y., Dou, Z., Zhang, M., Chang, Y., Zhao, X. (eds.) AIRS 2016. LNCS, vol. 9994, pp. 224–236. Springer, Cham (2016). doi:10.1007/978-3-319-48051-0_17

29. Zheng, Y., Tang, B., Ding, W., Zhou, H.: A neural autoregressive approach to collaborative filtering. arXiv preprint arXiv:1605.09477 (2016)

Data Mining for IR

Entity Ranking by Learning and Inferring Pairwise Preferences from User Reviews

Shinryo Uchida[1], Takehiro Yamamoto[1(✉)], Makoto P. Kato[1],
Hiroaki Ohshima[2], and Katsumi Tanaka[1]

[1] Graduate School of Informatics, Kyoto University, Kyoto, Japan
{uchida,tyamamot,kato}@dl.kuis.kyoto-u.ac.jp,
tanaka.katsumi.85e@st.kyoto-u.ac.jp
[2] Graduate School of Applied Informatics, University of Hyogo, Kobe, Japan
ohshima@ai.u-hyogo.ac.jp

Abstract. In this paper, we propose a method of ranking entities (e.g. products) based on pairwise preferences learned and inferred from user reviews. Our proposed method finds expressions from user reviews that indicate pairwise preferences of entities in terms of a certain attribute, and learns a function that determines the relative degree of the attribute to rank entities. Since there are a limited number of such expressions in reviews, we further propose a method of inferring pairwise preferences based on attribute dependencies obtained from reviews. As some pairwise preferences are less confident, we also propose a modified version of a learning to rank method, Fuzzy Ranking SVM, which can take into account the uncertainty of pairwise preferences. The experiment was carried out with three categories of products and several attributes specific to each category. The experimental results showed that our approach could learn more accurate pairwise preferences than baseline methods, and inference based on the attribute dependency could improve the performances.

Keywords: Entity ranking · Learning to rank · Review mining

1 Introduction

People often evaluate products and services in terms of a wide variety of their attributes. Some attributes such as price and size can be evaluated without use of the entity, while other attributes such as usability and portability cannot be evaluated without use of the entity. We call the former *search attributes* and the latter *experience attributes* by following the classification of goods [20].

It would be useful for product or service search systems to allow users to input desired experience attributes, since users usually have information needs based on product or service experiences. For example, when people want to buy a camera, they often have some reasons based on experiences that they expect (*e.g.* high portability for hiking or high photo quality for capturing beautiful scenery). In addition, as it is sometimes difficult to estimate experience attributes based

© Springer International Publishing AG 2017
W.-K. Sung et al. (Eds.): AIRS 2017, LNCS 10648, pp. 141–153, 2017.
https://doi.org/10.1007/978-3-319-70145-5_11

on search attributes, search by experience attributes is helpful especially for non-expert users. For example, it is hard for non-experts to judge which sensors can take more beautiful pictures: an APS-C sensor or an APS-H sensor.

In this paper, we propose a method of ranking entities based on the degree of a given experience attribute that is estimated by search attributes. There are several challenges to realize such entity rankings. First, while search attributes are objective and explicitly described on official Web sites, experience attributes are, by definition, subjective and only described in user reviews. Second, the degree of an experience attribute does not always correlate to the number of reviews in which the experience attribute is mentioned. For example, camera A can be frequently referred to as "a light camera" because of not its lightness but its popularity. Third, the evaluation for experience attributes is not absolute but often relative. Sentence "Camera A is stable" may not indicate high stability in general, but can only indicate high stability among a particular group of cameras. Therefore, simple aggregation of user reviews may not be effective for estimating experience attributes.

To address those challenges in realizing entity ranking based on experience attributes, we propose a method of estimating the relative value of each experience attribute by leveraging pairwise preferences in user reviews. We assume that experience attributes are always described by comparison with the other entities. Thus, we do not rely on opinions in reviews in which compared entities are unclear, but use only opinions that clearly describe what are compared. Such opinions are often expressed as comparative sentences, and can be modeled as pairwise preferences in terms of a certain attribute. For example, sentence "Camera A is more stable than camera B" indicates that camera A is larger than camera B in terms of the stability. We use such pairwise preferences to learn a function that determines the relative value of entities based on search attributes.

Since there are a limited number of comparative sentences in user reviews, we further propose a method of inferring pairwise preferences based on *attribute dependencies* mined from user reviews. For example, we can obtain an attribute dependency between the weight and stability from sentence "This camera is stable since it is heavy", and infer that camera A is more stable than camera B if we have already known that camera A is heavier than camera B. As some pairwise preferences, in particular the inferred preferences, are less confident, we also propose a modified version of a learning to rank method, *Fuzzy Ranking SVM*, which can take into account the uncertainty of pairwise preferences.

The experiments were carried out with three types of products. We examined whether the estimated rankings could concur with ones extracted from user reviews. The experimental results showed that rankings based on pairwise preferences were more accurate than baselines relying on the frequency of experience attributes in user reviews, and that inference based on the attribute dependency could enhance the accuracy of rankings.

The contributions of this paper are summarized as follows: (1) we proposed a method of estimating the relative value of experience attributes by leveraging pairwise preferences and attribute dependencies mined from user reviews;

(2) we introduce a novel machine learning method, Fuzzy Ranking SVM, which is the combination of Fuzzy SVM [18] and Ranking SVM [15], to deal with unconfident pairwise preferences; and (3) we conducted experiments and revealed high performances of the proposed methods in product domains.

The remainder of this paper is organized as follows. Section 2 surveys related work on review mining and entity ranking. Section 3 introduces the details of our proposed method. Section 4 presents experimental settings and results. Finally, Sect. 5 concludes by outlining future work.

2   Related Work

This section introduces related work on review mining and entity ranking.

2.1   Review Mining

One type of work utilizing user reviews is the summarization for providing the overview of all reviews. Hu and Liu proposed a mining and summarizing method for customers' opinions written on reviews [12]. They performed association mining to find the correspondence between entity attributes and customers' evaluations. Liu *et al.* proposed a supervised pattern discovery method for customers' opinions and a visualization method for the mined opinions [19]. Carenini *et al.* introduced a mining method from free-form texts [3]. Their system organizes entity attributes mined from documents in a hierarchical structure.

Another type of work utilizing user reviews is the aggregation for directly ranking entities. Choi *et al.* introduced a consensus search problem, which was defined as the problem of ranking entities based on user votes, and proposed a ranking method based on user consensus [6]. Their ranking method aggregates comments referring to an entity as votes from users. Ganesan *et al.* also proposed a ranking method based on how well opinions on entities match users' preferences [11].

The existing work focuses on the relationship between entity attributes and polarity (positive or negative), and uses reviews as surrogates of entities in search, while our work tackles the problem of learning the relationship between search and experience attributes by mining user reviews, and uses reviews to quantify each experience attribute for entity ranking.

2.2   Entity Ranking

Entity ranking has been addressed mainly in some tracks in TREC and INEX. The TREC Enterprise track proposed expert search tasks where participants were asked to create a ranking of people who are experts in a given topic [1,2,7,21]. The INEX Entity Ranking track prepared two tasks: entity ranking and entity list completion tasks [8–10]. The entity ranking task expects systems to return relevant entities in response to a given query, while the entity list completion task expects systems to return entities related to given example entities.

Several work has tackled the problem of finding related entities derived from user comparison. Jindal and Liu proposed a method of finding entity pairs compared in documents [14], and applied it to the related entity finding task. Li *et al.* proposed a weakly-supervised method for comparative question identification and comparable entity extraction [17].

There are some studies of ranking entities based on user opinions or evaluations by crowd-sourcing services. Chen *et al.* proposed a model to predict a ranking based on pairwise comparisons via crowdsourcing [4]. Cheng *et al.* studied the problem of entity ranking allowing predictions to be in the form of partial orders instead of total orders [5]. Kim *et al.* proposed a probabilistic model for learning multiple latent rankings using pairwise comparisons [16].

Iwanari *et al.* tackled a problem of ordering entities in terms of a given adjective by utilizing some evidences extracted from texts [13]. Their task is similar to ours as both work addresses entity ranking in terms of a certain ordering criterion. While their method utilizes description about entities for ranking, our method utilizes attributes of entities for ranking through a function learned by description about entities.

3   Methodology

In this section, we propose a method of estimating the relative value of each experience attribute by leveraging pairwise preferences in user reviews. Our proposed method consists of the following three steps:

1. Extracting pairwise preferences and attribute dependencies from user reviews;
2. Inferring additional pairwise preferences by the extracted pairwise preferences and attribute dependencies; and
3. Learning to rank entities by using the pairwise preferences as training data.

Each of the following subsections explains each step mentioned above.

3.1   Extracting Pairwise Preferences and Attribute Dependencies

Pairwise Preference. One of two entities can be high or low in terms of a certain attribute. We call this relation *comparative relation* on a particular attribute, and formally define it as binary relation $\prec_a$ on a set of entities E, where a is an attribute (either a search or an experience attribute). When the attribute value of a for e_j is larger than that for e_i, we write $e_i \prec_a e_j$ and comparative relation $\prec_a$ should include *pairwise preference* (e_i, e_j).

We estimate pairwise preferences in comparative relation $\prec_{a_k}$ for each attribute, by finding sentences matching predefined syntactic patterns. We first conduct morphological analysis and modification analysis of sentences in user reviews. Each sentence is modeled as a directed graph after these analyses, in which nodes are clauses composed of terms and edges are modification relations between two clauses. Syntactic patterns are also represented as a direct graph, in which nodes (or clauses) are represented as $\langle [x{:}p] \ \ldots \rangle$ representing term x

whose part of speech is p, and edges (or modification relations) are represented as $\rightarrow$. Some nodes in syntactic patterns contain terms with slots (denoted by variables) and terms matching the slots are extracted. For example, given sentence $\langle$[値段 (price):noun][が (is):particle]$\rangle \rightarrow \langle$[高い (high):adjective]$\rangle$, pattern $\langle$[x:noun][が(is):particle]$\rangle \rightarrow \langle$[y:adjective]$\rangle$ matches this sentence returning $x =$ "値段" (price) and $y =$ "高い" (high)[1].

Table 1. Extraction patterns for comparative relations for Japanese texts. English translation can be found in the parentheses.

subject	$\langle$[e_s:noun][の (of):particle]$\rangle \rightarrow \langle$[方 (more):noun][が (is):particle]$\rangle \rightarrow \langle$($a$:attribute)$\rangle$
object	$\langle$[e_c:noun][より (more):particle]$\rangle \rightarrow \langle$($a$:attribute)$\rangle$
	$\langle$[e_c:noun][と, に (with):particle]$\rangle \rightarrow \langle$[比べる (compare):verb]$\rangle \rightarrow \langle$($a$:attribute)$\rangle$
	$\langle$[e_c:noun][と (with):particle]$\rangle \rightarrow \langle$[比較する (compare):verb]$\rangle \rightarrow \langle$($a$:attribute)$\rangle$

Table 1 shows predefined syntactic patterns for comparative relations. The syntactic patterns consist of two parts: a subject part and an object part. A term matched with e_s in a subject part is extracted as the subject of the comparative sentence, while a term matched with e_o in an object part is extracted as the one compared with the subject. In this table, (a:attribute) represents an attribute expression that is defined as either (i) a adjective, (ii) a verb with adjectives "やすい" (easy) or "できる" (able), or (iii) a pair of a noun and an adjective. For example, attribute expressions can be portable, compact, and (photo, high-quality). As a result of the pattern matching, we can obtain entities e_s and e_o, and attribute a, and estimate that $e_o \prec_a e_s$ holds for pairwise preference (e_o, e_s). Since a subject or an object of comparative sentences are frequently omitted, we need to estimate missing subjects or objects. If a subject is missing, we assume that the subject is entity e for which the review was written only if the object is not e. In a similar way, if an object is missing, we assume that the object is entity e for which the review was written only if the subject is not e.

Each pairwise preference in a comparative relation is assigned with an initial confidence score based on the frequency in user reviews. Assuming that θ is the probability that a comparative sentence is incorrect, we define a confidence score as the probability that none of the comparative sentences regarding entity pair (e_i, e_j) and attribute a are correct:

$$\text{conf}_{\text{com}}((e_i, e_j), a) = 1 - (1 - \theta)^{n^a_{e_i, e_j}}, \tag{1}$$

where $n^a_{e_i, e_j}$ is the number of sentences that yield pairwise preference (e_i, e_j) and a as a result of the pattern matching. As more sentences contain a pairwise preference for an attribute, its confidence score becomes higher.

---

[1] Note that we use Japanese syntactic patterns in this paper since our experiments were conducted with Japanese reviews. English translation can be always found in the parentheses where Japanese terms are used in this paper.

Attribute Dependency. When a high value of attribute a implies a high (or low) value of attribute a' for any entities, we call it *attribute dependency* and write $a \rightarrow_+ a'$ (or $a \rightarrow_- a'$). As we mentioned earlier, this attribute dependency can be used for inferring additional pairwise preferences in comparative relations that cannot be found in user reviews.

We mine attribute dependencies from two clauses between which a resultative connection exists. Some conjunctive particles are useful indicators for resultative connections, for example, "ので" and "から", both of which mean "because" in English. More specifically, we used the following patterns to extract attribute dependencies: $\langle(a\text{:attribute})[$ので$\text{ (because):particle}]\rangle \rightarrow \langle(a'\text{:attribute})\rangle$ and $\langle(a\text{:attribute})[$から$\text{ (because):particle}]\rangle \rightarrow \langle(a'\text{:attribute})\rangle$. When a sentence matches one of the patterns, we expect that $a\rightarrow_- a'$ holds if only one of the attribute expressions is a negative clause; otherwise, we expect that $a\rightarrow_+ a'$ holds. For example, given sentence $\langle[$軽い$\text{ (light):adjective}][$ので$\text{ (because):particle}]\rangle \rightarrow \langle[$運ぶ$\text{ (carry):verb}][$やすい$\text{ (easy):adjective}]\rangle$, we obtain "軽い" (light) $\rightarrow_+$ "運" (carry).

In a similar way to the pairwise preference, we compute a confidence score for $a\rightarrow_+ a'$ (or $a\rightarrow_- a'$) as follows:

$$\text{conf}_{\text{dep}}((a, a')) = 1 - (1 - \phi)^{n_{a,a'}}, \tag{2}$$

where ϕ is a the probability that a resultative connection is correct, and $n_{a,a'}$ is the number of sentences yielding $a\rightarrow_+ a'$ (or $a\rightarrow_- a'$).

3.2 Inferring Pairwise Preferences Based on Attribute Dependencies

Since there are a limited number of pairwise preferences in comparative relations, we propose a method of inferring additional pairwise preferences based on attribute dependencies. We hypothesize that the attribute value of a' for entity e_j is larger than that for entity e_i if the attribute value of a for entity e_j is larger than that for entity e_i, and a high attribute value of a implies a high attribute value of a'. More formally, we define the following inference rules:

$$e_i \prec_a e_j \wedge a\rightarrow_+ a' \Rightarrow e_i \prec_{a'} e_j \tag{3}$$
$$e_i \prec_a e_j \wedge a\rightarrow_- a' \Rightarrow e_j \prec_{a'} e_i \tag{4}$$

For example, if we know that the weight of camera A is heavier than that of camera B and high weight implies high stability in cameras, we can infer that the stability of camera A is heavier than that of camera B.

The confidence score of pairwise preferences obtained by applying these inference rules can be computed as the product of the confidence scores of used pairwise preferences and attribute dependencies:

$$\text{conf}_{\text{com}}((e_i, e_j), a') = \text{conf}_{\text{com}}((e_i, e_j), a)\text{conf}_{\text{dep}}((a, a')) \tag{5}$$

Algorithm 1. Inferring pairwise preferences in comparative relations

---

Require: $G = (A, D)$
Ensure: R is a set of additional pairwise preferences in comparative relations
  $R \leftarrow \{\}$
  $P \leftarrow \{\}$ # a set of nodes that have been processed
  while $P \neq A$ **do**
    for all $a \in A$ that has no in-link **do**
      for all (a, a') for which $a \rightarrow_+ a'$ (or $a \rightarrow_- a'$) holds **do**
        for all (e_i, e_j) for which $e_i \prec_a e_j$ holds **do**
          append a new comparative relation $e_i \prec_a e_j$ (or $e_j \prec_a e_i$) to R
        end for
        $D \leftarrow D - \{(a, a')\}$
      end for
      append a to P
    end for
  end while

We can repeatedly apply the inference rules to infer more pairwise preferences, while different sequences of applications of the inference rules can result in different confidence scores. In this paper, we use an algorithm described in Algorithm 1, which can guarantee a unique result. Regarding each attribute as a node and each attribute dependency as an edge, we first construct a directed graph $G = (A, D)$, where A is a set of attributes and D is a set of attribute pairs for which the attribute dependency holds. We then look for node a with no in-link, apply the inference rules with attribute dependencies $a \rightarrow_+ a'$ or $a \rightarrow_- a'$, and remove edges corresponding to the used attribute dependencies. We repeat this process until all the nodes are processed. Note that this algorithm requires that directed graph G has no cycle. If G has cycles, we repeatedly remove elements with the lowest confidence score from D until all the cycles are removed.

3.3 Learning to Rank Entities by Using Pairwise Preferences

Having obtained pairwise preferences in comparative relations as a result of the review mining and applications of the inference rules, we learn a ranking function that uses search attributes for predicting the relative value of experience attributes. We first introduce a new ranking model for taking into account pairwise preferences with confidence scores, and then explain the application of the model to our problem.

Fuzzy Ranking Support Vector Machine. Fuzzy Ranking Support Vector Machine (Fuzzy Ranking SVM) is based on two SVM-based models: Ranking SVM and Fuzzy SVM. Ranking SVM was introduced by Joachim [15] and is an application of SVM for solving ranking problems. Ranking SVM uses pairwise preferences as training data to learn a ranking model that predicts the preference for a given pair. Fuzzy SVM was introduced by Lin *et al.* [18] and is an application of SVM for training data with weights. Fuzzy SVM gives different weights to input points for setting different contributions to the learning.

Fuzzy Ranking SVM, the combination of Ranking SVM and Fuzzy SVM, is formulated as follows:

$$\text{minimize} : \quad \frac{1}{2}\boldsymbol{w} \cdot \boldsymbol{w} + C \sum s_{i,j}\xi_{i,j} \qquad (6)$$

$$\text{subject to} : \quad \{\forall(\boldsymbol{x}_i, \boldsymbol{x}_j) \text{ s.t. } \boldsymbol{x}_i \prec \boldsymbol{x}_j\} : \boldsymbol{w}(\boldsymbol{x}_j - \boldsymbol{x}_i) \geq 1 - \xi_{i,j} \qquad (7)$$

$$\forall(i,j) : \xi_{i,j} \geq 0, \qquad (8)$$

where $\boldsymbol{w}$ is a weight vector to be learned, C is a constant parameter, $\xi_{i,j}$ is a slack variable, and $\boldsymbol{x}_i$ and $\boldsymbol{x}_j$ represent points in the training data. Binary relation $\prec$ is defined on points in the training data. A parameter $s_{i,j} \in (0,1]$ is an important component in Fuzzy SVM, which is the fuzzy membership of $\boldsymbol{x}_i$ and $\boldsymbol{x}_j$. A smaller value for $s_{i,j}$ reduces the effect of slack variable $\xi_{i,j}$, and, accordingly, the importance of $\boldsymbol{x}_i$ and $\boldsymbol{x}_j$ becomes smaller. Each point in the training data is a tuple of $((\boldsymbol{x}_i, \boldsymbol{x}_j), s_{i,j})$ such that $\boldsymbol{x}_i \prec \boldsymbol{x}_j$ holds.

Application of Fuzzy Ranking SVM to Entity Ranking. We finally explain the application of Fuzzy Ranking SVM to the entity ranking by experience attributes. First, we represent entities by their search attributes. Letting E be a set of entities and M be the number of search attributes of the entity set, each entity e_i can be represented by M-dimensional vector $\mathbf{e}_i$, in which a value for the k-th dimension is the attribute value of the k-th search attribute. The search attribute values are collected from their specification described on Web sites and are normalized into $[0,1]$.

Second, we construct training data required by Fuzzy Ranking SVM as follows. Having applied the review mining and inference described in Sects. 3.1 and 3.2, we obtain a set of pairwise preferences with confidence scores for each experience attribute a. Letting $\boldsymbol{x}_i = e_i$, $\boldsymbol{x}_j = e_j$, $s_{i,j} = \text{conf}_{\text{com}}((e_i, e_j), a)$ such that $e_i \prec_a e_j$ holds, we can build training data for each experience attribute a.

Finally, Fuzzy Ranking SVM is applied to the resultant training data and can learn function $f_a : \mathbb{R}^M \times \mathbb{R}^M \mapsto \{1, -1\}$ that indicates which entities is larger in terms of experience attribute a. Given experience attribute a, our proposed method can rank entities by using function f_a.

4   Experiments

In this study, we attempt to answer the following research questions by conducting the experiment: **RQ1** Does our proposed method improve the performance compared with the baselines that rely on the frequency of an experience attribute in user reviews?; **RQ2** Does our inference of pairwise preferences improve the the performance?; and **RQ3** Does Fuzzy Ranking SVM properly capture the uncertainty of pairwise preferences?

4.1   Experimental Setup

Dataset. We used Kakaku.com[2], which is a popular Japanese e-commerce site, to prepare the dataset for the experiment. We used three *categories* in the product domain, namely, *cameras*, *smartphones*, and *headphones*. For each category, we extracted entities, their search attributes, and user reviews from the Web sites. We chose the top 10 experience attributes that frequently appear in user reviews and used them in the experiment. Table 2 shows the statistic of the dataset and example experience attributes. "Dimension" in the table represents the dimensionality of vector $\mathbf{e}_i$ in each category.

Table 2. Statistics of categories used in experiment.

Category	# of Entities	# of Reviews	Dimension	Example experience attributes
Cameras	688	15,473	47	*usability, feeling of hold*
Smartphones	624	33,731	186	*easy-to-hold, comfort*
Headphones	2,229	13,117	304	*sound clearness, sound richness*

Baselines and Proposed Methods. To investigate the effectiveness of our proposed method, we prepared the following baselines relying on the frequency of an experience attribute in user reviews:

Frequency-based method (Freq). This method ranks entities in order of the frequency of an experience attribute in their user reviews. When an experience attribute frequently appears in user reviews of an entity, it is ranked higher.

Regression-based method (Reg). This method tries to learn a linear regression model, where the frequency of an experience attribute in user reviews for an object is a dependent variable and the values of search attributes of the object are explanatory variables.

We also compared the effectiveness of our proposed method with the following three conditions:

Proposed method without inference and confidence scores (Plain). This method does not infer pairwise preferences and simply uses the pairwise preferences extracted from user reviews. In addition, it does not compute confidence scores for pairwise preferences, *i.e.* confidence scores were always set to 1.0.

Proposed method without confidence scores (Infer). This method infers the additional pairwise preferences but does not compute confidence scores.

---

[2] http://kakaku.com.

Proposed method (Infer+Conf.) This method infers additional pairwise preferences and computes their confidence scores using Eqs. (1) and (5).

Evaluation Metric. We regarded the pairwise preferences written in the user reviews as the ground truth data, and examined whether a predicted preference could concur with them. We measured the accuracy of a method with the leave-one-out cross validation. More specifically, we removed one pairwise preference from the ground truth data and used the rest as the training data. The accuracy is computed as the fraction of the number of pairwise preferences predicted correctly among the ground truth data.

Implementation. We implemented Fuzzy Ranking SVM by extending the SVC class in scikit-learn[3], a machine learning library for Python. We used the RBF kernel for Fuzzy Ranking SVM and set their parameters with the default values of the SVC class. We used MeCab as a Japanese morphological analyzer and CaboCha as a Japanese modification analyzer. The parameters θ and ϕ in Eqs. (1) and (2) were empirically set to 0.5 and 0.1, respectively.

4.2 Results

Table 3 summarizes the accuracy of the baseline and proposed methods for three categories. We computed the accuracy for each experience attribute, and the macro average for 10 experience attributes is reported in the table.

As for **RQ1**, when we compare the baseline methods and Infer+Conf, we can see that Infer+Conf outperformed the baseline methods in terms of accuracy for all the three categories. This result suggests that the frequency of an experience attribute in user reviews is not always indicates the value of an experience attribute.

As for **RQ2**, by comparing the results of Plain and Infer, we can observe that Infer achieved the higher accuracy for *cameras* and *smartphones* categories, which indicates that our inference of pairwise preferences using attribute dependency could successfully increase the size of training data. However, we could not find such an improvement for the *headphones* category. To further investigate the effect of the inference, we analyzed the number of pairwise preferences inferred by the method. Table 4 shows the number of pairwise preferences before/after the inference. From the table, we can see that our inference method increased the number of pairwise preferences for the *cameras* and *smartphones* categories while few preferences were inferred in the *headphones* category, which correlates the performance improvement of Infer against Plain.

One reason of the little effect on the *headphones* category is our method failed to extract attribute dependencies and thus failed to infer the additional preferences. Our inferring method relies on sentences having resultative connections. However, we found that there were few such sentences in the *headphones*

---

[3] http://scikit-learn.org/.

Table 3. Accuracy for three categories. Highest values among methods are in bold.

Category	Baselines		Proposed		
	Freq	Reg	Plain	Infer	Infer+Conf
Cameras	.576	.499	.603	.688	**.705**
Smartphones	.524	.493	.486	.574	**.575**
Headphones	.498	.437	**.701**	**.701**	**.701**
Macro Average	.531	.476	.597	.654	**.660**

Table 4. Average number of pairwise preferences before/after inference.

Category	Before inference	After inference
Cameras	20.6	99.7
Smartphones	37.9	369.4
Headphones	9.9	10.1

category. One approach to solving this problem would be extracting attribute dependencies based on the *co-occurrence* of attributes in sentences rather than the resultative connections of attributes.

As for **RQ3**, we can see that Infer+Conf outperformed Infer for the *cameras* category while little improvement could be found on the *smartphones* and *headphones* categories. One reason for the different effects on *cameras* and *smartphones* categories of Infer+Conf would be accuracy of confidence scores for the pairwise preferences. If a method wrongly gives low confidence scores to the correct pairwise preferences and high scores to wrong pairwise preferences, the performance of the resultant function learned by Fuzzy Ranking SVM would not be improved. Exploring more precise methods for computing confidence scores for the pairwise preference is an interesting research direction of this study.

5 Conclusions

In this paper, we proposed a method of ranking entities in terms of an experience attribute based on their search attributes. Our method first extracts the pairwise preferences of entities from user reviews, and infers the additional pairwise preferences by using attribute dependencies extracted from user reviews. The extracted and inferred preferences were used to learn a ranking function by applying Fuzzy Ranking SVM, which can take into account the uncertainty of pairwise preferences.

The experimental results with three categories in product domain revealed that our proposed method, which predicts the ranking of entities by pairwise preferences, outperformed the baselines relying on the frequency of experience attributes in user reviews in terms of accuracy, and that the inference based on attribute dependency could improve the accuracies for two of three categories.

Acknowledgments. This work was supported by JSPS KAKENHI Grant Numbers JP15H01718, JP26700009, JP16H02906, JP16K16156, and JP25240050.

References

1. Bailey, P., De Vries, A.P., Craswell, N., Soboroff, I.: Overview of the TREC 2007 enterprise track. In: TREC (2007)
2. Balog, K., Thomas, P., Craswell, N., Soboroff, I., Bailey, P., De Vries, A.P.: Overview of the TREC 2008 enterprise track. In: TREC (2008)
3. Carenini, G., Ng, R.T., Zwart, E.: Extracting knowledge from evaluative text. In: K-CAP, pp. 11–18 (2005)
4. Chen, X., Bennett, P.N., Collins-Thompson, K., Horvitz, E.: Pairwise ranking aggregation in a crowdsourced setting. In: WSDM, pp. 193–202 (2013)
5. Cheng, W., Rademaker, M., De Baets, B., Hüllermeier, E.: Predicting partial orders: ranking with abstention. In: ECML-PKDD, pp. 215–230 (2010)
6. Choi, J., Kim, D., Kim, S., Lee, J., Lim, S., Lee, S., Kang, J.: Consento: a new framework for opinion based entity search and summarization. In: CIKM, pp. 1935–1939 (2012)
7. Craswell, N., de Vries, A.P., Soboroff, I.: Overview of the TREC 2005 enterprise track. In: TREC (2005)
8. de Vries, A.P., Vercoustre, A.-M., Thom, J.A., Craswell, N., Lalmas, M.: Overview of the INEX 2007 entity ranking track. In: Fuhr, N., Kamps, J., Lalmas, M., Trotman, A. (eds.) INEX 2007. LNCS, vol. 4862, pp. 245–251. Springer, Heidelberg (2008). doi:10.1007/978-3-540-85902-4_22
9. Demartini, G., Iofciu, T., de Vries, A.P.: Overview of the INEX 2009 entity ranking track. In: Geva, S., Kamps, J., Trotman, A. (eds.) INEX 2009. LNCS, vol. 6203, pp. 254–264. Springer, Heidelberg (2010). doi:10.1007/978-3-642-14556-8_26
10. Demartini, G., de Vries, A.P., Iofciu, T., Zhu, J.: Overview of the INEX 2008 entity ranking track. In: Geva, S., Kamps, J., Trotman, A. (eds.) INEX 2008. LNCS, vol. 5631, pp. 243–252. Springer, Heidelberg (2009). doi:10.1007/978-3-642-03761-0_25
11. Ganesan, K., Zhai, C.: Opinion-based entity ranking. Inf. Retrieval **15**(2), 116–150 (2012)
12. Hu, M., Liu, B.: Mining and summarizing customer reviews. In: KDD, pp. 168–177 (2004)
13. Iwanari, T., Yoshinaga, N., Kaji, N., Nishina, T., Toyoda, M., Kitsuregawa, M.: Ordering concepts based on common attribute intensity. In: IJCAI, pp. 3747–3753 (2016)
14. Jindal, N., Liu, B.: Identifying comparative sentences in text documents. In: SIGIR, pp. 244–251 (2006)
15. Joachims, T.: Optimizing search engines using click through data. In: KDD, pp. 133–142 (2002)
16. Kim, Y., Kim, W., Shim, K.: Latent ranking analysis using pairwise comparisons. In: ICDM, pp. 869–874 (2014)
17. Li, S., Lin, C.Y., Song, Y.I., Li, Z.: Comparable entity mining from comparative questions. IEEE Trans. Knowl. Data Eng. **25**(7), 1498–1509 (2013)
18. Lin, C.F., Wang, S.D.: Fuzzy support vector machines. IEEE Trans. Neural Networks **13**(2), 464–471 (2002)
19. Liu, B., Hu, M., Cheng, J.: Opinion observer: analyzing and comparing opinions on the web. In: WWW, pp. 342–351 (2005)

20. Nelson, P.: Information and consumer behavior. J. Polit. Econ. **78**(2), 311–329 (1970)
21. Soboroff, I., de Vries, A.P., Craswell, N.: Overview of the TREC 2006 enterprise track. In: TREC (2006)

Probabilistic Local Matrix Factorization Based on User Reviews

Xu Chen[1], Yongfeng Zhang[2], Wayne Xin Zhao[3($\boxtimes$)], Wenwen Ye[1], and Zheng Qin[1]

[1] School of Software, Tsinghua University, Beijing, China
[2] College of Information and Computer Sciences,
University of Massachusetts Amherst, Amherst, USA
[3] School of Information, Renmin University of China, Beijing, China
batmanfly@gmail.com

Abstract. Local matrix factorization (LMF) methods have been shown to yield competitive performance in rating prediction. The main idea is to leverage the ensemble of submatrices for better low-rank approximation. However, the generated submatrices and recommendation results in the existing methods are usually hard to interpret. To address this issue, we adopt a probabilistic approach to enhance model interpretability of LMF methods by leveraging user reviews. In specific, we incorporate item-topics to construct meaningful "local clusters", and further associate them with opinionated word-topics to explain the corresponding semantics and sentiments of users' ratings. The proposed approach is a joint model which characterizes both ratings and review text. Extensive experiments on real-world datasets demonstrate the effectiveness of our proposed model compared with several state-of-art methods. More importantly, the produced results provide meaningful explanations to understand users' ratings and sentiments.

1 Introduction

Recently, local matrix factorization (LMF) has attracted increasing attention [4,10,19] in recommender system community. LMF methods have been shown to give better performance than traditional matrix factorization (MF) techniques [9,16] in rating prediction. Typically, LMF methods identify subgroups of users and items, and construct multiple submatrices based on the original user-item rating matrix. They apply traditional MF methods to each submatrix individually, and combine the results from submatrices as the final prediction. Such an approach aims to enhance the low-rank property of submatrices and improve parallel processing.

So far, existing LMF methods mainly focus on how to "look for" subgroups using the proximity criterion, including random dividing [13], kernel smoothing [10] and Bregman co-clustering [6]. An important aspect has been usually ignored, *i.e.*, model interpretability. Due to this, several queries cannot be well answered by previous studies, including why such subgroups are formed, what are the semantics of each subgroup, how a user in a subgroup rates and comments. In

W.-K. Sung et al. (Eds.): AIRS 2017, LNCS 10648, pp. 154–166, 2017.
https://doi.org/10.1007/978-3-319-70145-5_12

addition to performance, these problems are fundamental to understand users' rating behaviors and explain recommendation results. It is important to consider these factors in order to design an effective recommender.

To address this issue, we propose a novel explainable probabilistic LMF model by leveraging user reviews. The key to explainable LMF is how to derive explainable subgroups. Inspired by [19], we adopt topic modeling to characterize items clusters as *item-topics*, and further assign users to these item-topics "softly" (*i.e.*, in a probabilistic way). An item-topic is essentially a multinomial distribution over the set of items and tends to cluster items with similar functions or categories, which is relatively easy to interpret. For each item-topic, we set topic-specific latent factors for both users and items. In this way, we can better understand users' rating behaviors with the help of topical contexts. Such a formulation partially provides the semantics of subgroups, however, it is still unable to explain why a user gives a high or low rating to an item. Our solution is to further leverage review information to further enhance the interpretability of prediction results. In specific, we incorporate opinionated word-topics like that in topic models [3] to characterize the semantics and sentiments reflected in review text. In our model, an item-topic is associated with a unique distribution over word-topics in each sentiment level, and the generation of review text is based on both item-topic and sentiment level. The incorporation of opinionated word-topics can improve the learning of item-topics, since review text is likely to contain relevant aspect words. In addition, we can identify the most associated words to explain the opinion polarity of user ratings.

To evaluate the performance of the proposed model, we conduct extensive experiments on real-world datasets. The experimental results indicate the effectiveness of our model. Especially, it has been shown to give better explainability for the learned models and prediction results. The main contribution of this work is to incorporate item-topics to construct meaningful subgroups, and associate them with opinionated word-topics mined from review text to explain the corresponding semantics and sentiments for users ratings. By using topic-specific latent factors, our model yields competitive performance while the learned item- and word-topics give good interpretability. To our knowledge, it is the first time that word-topics discovered in the review text have been utilized to explain LMF methods.

2   Related Work

Local matrix factorization. Recently, local matrix factorization (LMF) has received much attention [4,10,13,25,26], which aim to enhance the low-rank property and parallel processing. Typically, these methods split the original matrix into smaller submatrices, and then apply traditional MF technqiues [9,16] on submatrices individually. The final predictions are generated by combing the predictions from submatrices. In specific, the DFC model [13] randomly divided the original matrix into small subgroups; the kernel smoothing method was used to find nearest neighbors [10]; the Bregman co-clustering method was exploited to split the original matrix [4]. The most recent study [19] adopted a probabilistic approach to generating "soft" clusters, however, it cannot model review text.

Review-based recommendation. Review information has been shown to improve the performance of rating prediction [1,7,8,11,12,14,22,23]. The major benefits gained from review information can be summarized in two aspects. First, the user-item rating matrix is sparse, and the auxiliary textual information is able to alleviate this issue to some extent [21]. More importantly, textual contents in user reviews can provide explainable information for users' ratings [5,14,20,24]. The HFT model [14] directly transformed the latent factors in MF side into the topic distributions. By aligning latent factors with topics, the produced results help understand users' ratings. The PACO model [20] designed a poisson additive co-clustering model to build interpretable recommendation system. The EFM model [24] extracted product features and users' corresponding sentiments on them, and further used them to explain user ratings.

Our work is closely related to these studies, and makes a meaningful connection between LMF and explainable MF methods. Although the superiority of LMF methods in rating prediction has been shown, the explainability has seldom been well addressed. Hence, the semantics of each subgroup were not clear and users' rating behaviors cannot be well understood. As a comparison, we propose to use item-topics to explain subgroups and opinionated word-topics from review text to explain the ratings and sentiments.

3 Probabilistic Local Matrix Factorization Based on User Review

In this section, we present the proposed probabilistic local matrix factorization. To make more clear presentation, we first list the notations used throughout the paper in Table 1.

Table 1. Notations and descriptions.

Notations	Descriptions
N, M	number of rows (users) and columns (items)
K_1, K_2	the number of item- and word-topics respectively
D	the number of dimensions for latent vectors
u, i, c, k	index variables respectively for users and items, item- and word-topics
$\boldsymbol{p}_u^{(c)}, \boldsymbol{q}_i^{(c)}$	the latent vector ($\in \mathbb{R}^D$) respectively for user u and item i w.r.t. the item-topic c
$r_{u,i}/\boldsymbol{w}_{u,i}$	the rating and review text of user u on item i
$y_{u,i}$	item-topic assignment associated with $r_{u,i}$
$z_{u,i,j}$	word-topic assignment for the j-th word in $\boldsymbol{w}_{i,j}$
ψ^k	word distribution in word-topic k
θ^c	word-topic distribution of item-topic c
φ^c	item distribution in item-topic c
ϕ^u	item-topic distribution of user u
$\lambda_0, \lambda_P, \lambda_Q$	priors of latent factors
$\beta/\beta', \alpha/\alpha'$	priors of the word/item-topics and distributions over word/item-topics respectively

3.1   The Proposed Model

As indicated in [19], lack of interpretability has been one major issue for previous LMF methods [10,13]. These models cannot answer two typical queries well: (1) what are the semantics for the generated submatrices, and (2) why a user likes or dislikes an item. In our model, we aim to solve the above issues by considering two aspects. First, we adopt a probabilistic topic modeling approach to learn "soft" subgroup of the items, and each item-topic together with the associated ratings can be considered as a local (*i.e.*, topic-specific) view of the entire user-item rating matrix. Second, we incorporate opinionated word-topics to enhance the learning of item-topics, and describe how a user comment on an item with some sentiment. The final model integrates both aspects (*i.e.*, ratings and review).

Modeling local subgroups with probabilistic topic models. The key step for LMF methods lies in that how to generate subgroups of users and items, which will further form a corresponding submatrix. We first adopt item-topics to model item subgroups. Formally, an item-topic c is a multinomial distribution over the set of all the items, denoted by φ^c. Each entry φ_i^c denotes the probability of item i in item-topic c. Further, we model a user u's interests by a multinomial distribution over item-topics denoted by ϕ^u. Each entry ϕ_c^u denotes the probability that a user is likely to rate an item in item-topic c. The interest distribution can be considered as users' membership over item subgroups, which forms probabilistic user subgroups. Next, we study how to generate a rating triplet $\langle u, i, r_{u,i} \rangle$. In LMF methods, each user (or item) will be associated with a unique latent factor in different submatrices. In our case, an item-topic corresponds to a local submatrix. Following [19], we propose to use topic-specific latent factors. For each item-topic c, we set a corresponding latent factor $\boldsymbol{p}_u^c$ ($\boldsymbol{q}_i^c$) for user u (item i). When a user u starts to rate an item i, she first draw a topic assignment $y_{u,i}$ according to ϕ^u, and then generates item i using φ^c like that in LDA [3]. Once the item-topic assignment has been sampled, following PMF [16,19], the rating $r_{u,i}$ is generated by using a topic-specific Gaussian distribution

$$\mathcal{N}(r_{ui}|(\boldsymbol{p}_u^{y_{u,i}})^\top \cdot (\boldsymbol{q}_i^{y_{u,i}}), \sigma_{y_{u,i}}^2), \tag{1}$$

where $\sigma_{y_{u,i}}^2$ is the variance corresponding to topic $y_{u,i}$.

Modeling user reviews to explain the item subgroups. Above, a subgroup of items is modeled as an item-topic, which is a soft clustering of the items. These item-topics provide important topical contexts to explain rating predictions using topic-specific latent factors. However, the item-topics mainly reflect co-occurrence patterns based on users' rating history, and they cannot capture the sentiment level of a user towards an item, *i.e.*, why a user gives a high rating or a low rating. Intuitively, the opinion polarity of a user review tends to be more positive if her rating is higher, and the generation of review text is closely related to the sentiment levels of a user on an item. Let $\mathcal{O} = \{1, ..., l, ..., L\}$ be a set of L sentiment labels, in which each label l denotes a sentiment level and a higher level indicates a more positive polarity. Our key idea lies in that the generation of a review text should be based on both item-topic and sentiment level. Formally, we assume that there are a set of K_2 word-topics. Each word-topic k is modeled as a multinomial distribution over the terms in the vocabulary, denoted by ψ^k. We assume that an item-topic c will correspond to an opinionated distribution over word-topics for each sentiment level l, denoted by $\theta^{c,l}$. Each entry $\theta_k^{c,l}$ denotes the probability of word-topic k for item-topic c with sentiment label l. Let $\boldsymbol{w}_{u,i}$ denote a vector of words in the review associated with the rating record $\langle u, i, r_{u,i} \rangle$. For each word token $w_{u,i,j} \in \boldsymbol{w}_{u,i}$, we first draw a word-topic assignment $z_{u,i,j}$ according to $\theta^{z_{u,i},l_{u,i}}$, where $z_{u,i}$ and $l_{u,i}$ correspond to the item-topic and sentiment label respectively for $\langle u, i, r_{u,i} \rangle$. Then we generate word $w_{u,i,j}$ according to the word-topic $\psi^{z_{u,i,j}}$. Our generation process involves the sentiment label for a user review. There can be several ways to set the sentiment labels. Here we adopt a simple yet effective method: we consider two sentiment levels (*i.e.*, positive and negative) and set it based on the corresponding rating score. In a five-star rating system, the sentiment label of a user review is set to positive, if the rating score is higher than three stars, otherwise it will be set to negative.

The final model. Our final model combines the above two parts: it characterizes user ratings by probabilistic LMF and models user reviews to explain the item subgroups learned in LMF. We implement a full Bayesian formulation for this model. In specific, for each item-topic c, we put priors on topic-specific latent factors $\boldsymbol{p}_u^c$ and $\boldsymbol{q}_i^c$, denoted by $\boldsymbol{\lambda}_P^c, \boldsymbol{\lambda}_Q^c, \boldsymbol{\lambda}_0^c$ as in BPMF [17]; we also put priors on variables $\varphi^c, \phi^u, \psi^k$ and θ^c, denoted by $\alpha, \alpha', \beta, \beta'$ respectively. We refer to the proposed model as **ELMF** (*Explainable LMF*). The generation process and complete plate notation for our model have been shown in Figs. 1 and 2 respectively. In our model, the item-topic $y_{u,i}$ is not only used to generate the items, but associated with distributions over word-topics to generate review text. The incorporation of review text is able to improve the coherence of item subgroups (*i.e.*, topics), and also enhance the semantic explainability of user ratings. When generating review text, the sentiment label of a review also plays an important role. As indicated [18], when a user is praising or criticizing an item, the word topics she selects are likely to be different. Based on this consideration, the generation of review text is based on both item-topic and sentiment level.

Insights into the model interpretability of ELMF. The model interpretability of ELMF has been reflected in two aspects. First, it tries to look for more meaningful subgroups of items and users in a probabilistic way. We achieve this by capturing item co-occurrence patterns in rating history. The derived item-topics can be meaningful with similar functions or categories. For each item-topic, we set topic-specific latent factors for both users and items. In this way, we can better understand users' rating preference with the help of topical contexts. Second, an item-topic is associated with different word-topics in different sentiment labels. We incorporate textual contexts as opinionated word-topics to explain why a user likes or dislikes an item. The combination of these two aspects yields a better model explainability and meanwhile keeps the performance superiority of LMF methods.

1. For each item-topic $c = 1, ..., K_1$, draw a multinomial distribution over all the items $\varphi^c \sim Dir(\beta')$;
2. For each word-topic $k = 1, ..., K_2$, draw a multinomial distribution over all the words $\psi^k \sim Dir(\beta)$;
3. For item-topic $c = 1, ..., K_1$,
   i. Draw the hyperparameters of the user and item latent vectors $P(\lambda_P^c | \lambda_0^c)$ and $P(\lambda_Q^c | \lambda_0^c)$
   ii. For each item $i = 1, ..., M$, draw the topic-specific item latent vector $q_i^c \sim P(q_i^c | \lambda_Q^c)$;
   iii. For each user $u = 1, ..., N$, draw the topic-specific item latent vector $p_u^c \sim P(p_u^c | \lambda_P^c)$;
   iv. For each sentiment label $l = 1, ..., L$, draw a multinomial distribution over all the word-topics $\theta^{c,l} \sim Dir(\alpha)$;
3. For each user $u = 1, ..., N$,
   i. Draw a multinomial distribution over all the item-topics $\phi^u \sim Dir(\alpha')$;
   ii. For each rated item i by u,
      (1) Draw an item-topic $y_{u,i} \sim Disc(\phi^u)$;
      (2) Draw the item $i \sim Disc(\varphi^{y_{u,i}})$;
      (3) Draw the rating $r_{u,i} \sim \mathcal{N}(r_{u,i} | (p_u^{y_{u,i}})^\top \cdot (q_i^{y_{u,i}}), \sigma_{y_{u,i}}^2)$;
      (4) Set the sentiment label $l_{u,i}$ based on $r_{u,i}$;
      (5) For each word token $w_{u,i,j}$ in $w_{u,i}$,
         · Draw a word-topic $z_{u,i,j} \sim Disc(\theta^{y_{u,i},l_{u,i}})$;
         · Draw the word $w_{u,i,j} \sim Disc(\psi^{z_{u,i,j}})$.

Fig. 1. The generative process of the ELMF model.

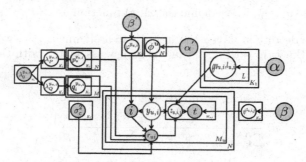

Fig. 2. The plate notation for our ELMF model.

3.2   Model Learning

We would like to learn the following parameters or variables: $\{\theta, \psi, \phi, \varphi, p, q\}$ by fixing the hyper-parameters. We aim to maximize the joint likelihood of observed ratings and review text. The problem is hard to directly optimize, and we adopt a collapsed Gibbs sampling method for solving it. Our learning tasks involve two major parts, inferring word- and item-topic assignments $\{y, z\}$ and optimizing the latent factors $\{p, q\}$. Once the topic assignments $\{y, z\}$ have been obtained, the distribution parameters $\{\theta, \psi, \phi, \varphi\}$ can be then estimated based on the word- and item-topic assignments. Let r, m, w, l, y and z be the vectors for ratings, items, words, sentiment labels, item-topic assignments and word-topic assignments respectively. Next we give the Gibbs sampling formula for topic assignments and the update formula for latent factors. For convenience, let Ψ denote all the hyper-parameters.

Sampling item-topics. Fixing all latent factors $\{\boldsymbol{p}, \boldsymbol{q}\}$, the item-topic assignment for the rating triplet $\langle u, i, r_{u,c}\rangle$ can be drawn according to:

$$P(y_{u,i} = c | \boldsymbol{r}, \boldsymbol{m}, \boldsymbol{w}, \boldsymbol{l}, \boldsymbol{y}_{\neg(u,i)}, \boldsymbol{z}, \boldsymbol{p}, \boldsymbol{q}; \Psi) \qquad (2)$$

$$\propto \frac{n_c^u + \alpha'}{\sum_{c'=1}^{K_1} n_{c'}^u + \alpha'} \times \frac{n_i^c + \beta'}{\sum_{i'=1}^{M} n_{i'}^c + \beta'}$$

$$\times \mathcal{N}(r_{u,i} | \boldsymbol{p}_u^c, \boldsymbol{q}_i^c, \sigma_c^2) \times \frac{\Delta(\boldsymbol{n}^{c,l_u,i} + \boldsymbol{n}^{u,i} + \boldsymbol{\alpha})}{\Delta(\boldsymbol{n}^{c,l_u,i} + \boldsymbol{\alpha})},$$

where n_c^u denotes the number of the rated items by user u assigned to item-topic c, n_i^c denotes the number that item i is assigned to item-topic c, $\boldsymbol{n}^{c,l_u,i}$ is a V-dimensional (vocabulary size) count vector in which $n_w^{c,l_u,i}$ denotes the number of word w attached to items in item-topic c with sentiment label $l_{u,i}$, $\boldsymbol{n}^{u,i}$ is a V-dimensional count vector in which $n_w^{u,i}$ denotes the number of word w appearing the review associated with rating $r_{u,i}$, and $\boldsymbol{\alpha}$ is a V-dimensional vector of equal value α. All the count statistics are computed by excluding the information associated with $\langle u, i, r_{u,i}\rangle$. We define the $\Delta(\boldsymbol{x})$ function as $\Delta(\boldsymbol{x}) = \frac{\Pi_{w=1}^V \Gamma(x_w)}{\Gamma(\sum_{w'=1}^V x_{w'})}$.

Sampling word-topics. Given the item-topic assignment $y_{u,i}$, we sample the word-topic for the j-th word $n_{w_{u,i,j}}^k$ in the review associated with the rating $r_{u,i}$ according to:

$$P(z_{u,i,j} = k | \boldsymbol{r}, \boldsymbol{m}, \boldsymbol{w}, \boldsymbol{y}, \boldsymbol{l}, \boldsymbol{z}_{\neg(u,i,j)}, \boldsymbol{p}, \boldsymbol{q}; \Psi)$$

$$\propto \frac{n_k^{y_{u,i},l_{u,i}} + \alpha}{\sum_{k'=1}^{K_2} n_{k'}^{y_{u,i},l_{u,i}} + \alpha} \times \frac{n_{w_{u,i,j}}^k + \beta}{\sum_{w'} n_{w'}^k + \beta}, \qquad (3)$$

where $n_k^{y_{u,i},l_{u,i}}$ is the number that words assigned to word-topic k with the associated item-topic $y_{u,i}$ and sentiment label $l_{u,i}$, $n_{w_{u,i,j}}^k$ denotes the number that word $w_{u,i,j}$ is assigned to word-topic k.

Updating latent factors. Following BPMF [17], given an item-topic c, $\boldsymbol{\lambda}_P^c = \{\mu_p^c, \Lambda_P^c\}$ is first generated according to a Gaussian-Wishart distribution with parameter $\boldsymbol{\lambda}_0^c = \{\mu_0^c, \nu_0^c, \boldsymbol{W}_0^c\}$ by $P(\boldsymbol{\lambda}_P^c | \boldsymbol{\lambda}_0^c) = \mathcal{N}(\mu_p^c | \mu_0^c, (\Lambda_P^c)^{-1}) \mathcal{W}(\Lambda_P^c | \nu_0^c, \boldsymbol{W}_0^c)$, then the conditional distribution over user u's latent factor $\boldsymbol{p}_u^c$ is:

$$P(\boldsymbol{p}_u^c | \boldsymbol{r}, \boldsymbol{q}^c; \Phi)$$

$$= \mathcal{N}(\boldsymbol{p}_u^c | \bar{\mu}_P^c, (\bar{\Lambda}_P^c)^{-1})$$

$$\propto P(\boldsymbol{p}_u^c | \mu_P^c, (\Lambda_P^c)^{-1}) \times \prod_{v=1}^{M} \mathcal{N}(r_{ui} | \boldsymbol{P}_u^c, \boldsymbol{q}_i^c, \sigma_c^2) \qquad (4)$$

where we have: $\bar{\Lambda}_P^c = \Lambda_P^c + \frac{1}{\sigma_c^2}\sum_{v=1}^M (\boldsymbol{q}_i^c(\boldsymbol{q}_i^c)^\top)^{\mathbb{I}(y_{u,i},c)}$ and $\bar{\mu}_P^c = (\bar{\Lambda}_P^c)^{-1}(\Lambda_P^c\mu_P^c +$
$\frac{1}{\sigma_c^2}\sum_{v=1}^M (\boldsymbol{q}_i^c r_{u,i})^{\mathbb{I}(y_{u,i},c)})$, where $\mathbb{I}(y_{u,i},c)$ is an indicator function which returns 1
only when $y_{u,i} = c$ otherwise 0. Following Eq. 5, $\boldsymbol{q}_i^c$ can be updated in a similar
way, we omit it here.

Computational complexity. Let D be the dimension number of latent fac-
tors in $\boldsymbol{p}_u^c$ and $\boldsymbol{q}_v^c$, $S = \sum_{u,c,i}\mathbb{I}(y_{u,i},c)$, K_1 be the number of item-topics,
then according to [19], updating users' and items' latent factors in each iter-
ation takes a cost of $\mathcal{O}(D^2 S + D^3 K_1 N + D^3 K_1 M)$. Let A be the number
of all the observed ratings, B be average number of words in a review, K_2
be the number of word-topics. The complexity for item-topic and word-topic
assignment is $\mathcal{O}(A(K_1 DB + BK_2))$. Hence, the overall cost in an iteration is
$\mathcal{O}(D^2 S + D^3 K_1 N + D^3 K_1 M + ABK_1 D + ABK_2)$

Rating Prediction. Once the model parameters have been learned, we can
predict the final ratings by: $\hat{r}_{u,i} = \sum_{c=1}^{K_1}\{\frac{Z_c}{Z_{(\cdot)}}(\boldsymbol{p}_u^c)^\top \cdot \boldsymbol{q}_i^c\}$, where $Z_c = \phi_c^u \times \varphi_i^c$
and $Z_{(\cdot)} = \sum_{c=1}^{K_1} Z_c$. Z_c can be understood as the weight coefficient for the
predictions from the item-topic c. Z_c will be large if both u and i have close
associations with item-topic c, i.e., ϕ_c^u and φ_i^c are large.

4   Experiments

In this section, we conduct evaluation experiments to examine the performance
of the proposed model.

4.1   Experimental Setup

Datasets. Six datasets [15] from diverse Amazon categories have been used as
evaluation collection. We present the statistics of the datasets in Table 2. For
each dataset, we randomly select 80% ratings to train our model, and the rest
are held out for testing.

Baseline methods. We compare our models with the following baselines:

- *PMF* [16]: It's a Bayesian probabilistic implementation of the traditional
  matrix factorization, shown to give accurate predictions in practice.
- *HFT* [14]: It's a competitive review-based rating prediction method by leveraging review information to enhance the prediction performance.
- *BPMTMF* [19]: It's a state-of-art local matrix factorization method which
  adopts a Bayesian formulation approach to model user ratings.

There can be more LMF baselines to compare here, including DFC [13],
LLORMA [10] and WEMAREC [4]. As shown in [19], BPMTMF is better than
these baselines. Our empirical results have also confirmed this. Hence, we only
select BPMTMF as the only representative for LMF baselines. We use a five-fold
cross validation to obtain the final performance for all the comparison methods.
To set various parameters in both baselines and our models, we use a grid search

Table 2. Statistics of our datasets.

Datasets	#Users	#Items	#Ratings	$\frac{\#Reviews}{\#Users}$	Density
Music	1492	900	7931	5.55	0.59%
Auto	2928	1835	18308	6.25	0.34%
Office	4905	2420	39974	8.14	0.33%
Patio	1686	962	11740	6.96	0.72%
Video	5130	1685	33146	6.46	0.38%
DigiMu	5541	3568	48255	8.71	0.24%

method by finding the values which lead to the best performance in five-fold cross validation. For our model ELMF, we set the number of latent factors to 20, the number of item-topics K_1 to 20, the number of word-topics K_2 to 15, and the hyper-parameters are set as follows $\alpha' = 0.01, \beta' = 2.5, \alpha = 0.1, \beta = 2.5, \mu_0^c = 0, \nu_0^c = 20, W_0^c$ is a identity matrix, $\sigma = \sqrt{2}$.

4.2 Evaluation on Rating Prediction

We first evaluate the performance of the proposed models in rating prediction, and adopt the commonly used Root Mean Square Error (RMSE) as the evaluation metrics, defined as $\sqrt{\frac{\sum_{\langle u,i,r_{u,i}\rangle \in \mathcal{D}} (r_{u,i} - \hat{r}_{u,i})^2}{|\mathcal{D}|}}$, where $|\mathcal{D}|$ is the number of samples in the test set $\mathcal{D}$.

Results and Analysis. We present the comparison results in Table 3 and have the following observations. First, HFT gives better performance than PMF in four of six datasets. HFT leverages review information to enhance rating prediction, which indicates the usefulness of review information. Second, BPMTMF performs best among the three baselines. The improvement margins over PMF and HFT are substantial, especially it yields 14.1% reduction of RMSE on the *DigiMu* dataset. It indicates the effectiveness of LMF methods for rating prediction. Third, our proposed models achieve similar or better performance compared with BPMTMF. As shown in [19], so far, BPMTMF has been the most competitive LMF method in rating prediction by comparing with DFC [13], LLORMA [10] and WEMAREC [4]. The margins that our models improve over BPMTMF are not large, while small RMSE reductions are likely to yield significant system performance improvement in practice [2]. The above results and analysis have shown the effectiveness of our proposed models.

Parameter Tuning. In our models, two important parameters are the number of item-topics (*i.e.*, K_1) and the number of word-topics (*i.e.*, K_2). Here, we examine how they affect the model performance. We select the *Music* dataset to report the tuning results and the rest datasets give the similar findings. We first find the optimal values by using cross validation. Then we fix one and vary the other. To tune K_1 and K_2, we use a grid search method by varying the values from 5 to 40 with a gap of 5. We present the tuning results in Fig. 3. Overall,

Table 3. Performance comparisons of RMSE results on six datasets. Smaller is better.

Datasets	PMF	HFT	BPMTMF	ELMF
Music	0.953	0.954	0.905	**0.902**
Auto	1.004	0.981	0.970	**0.969**
Office	0.955	0.965	0.950	**0.947**
Patio	1.110	1.091	1.070	**1.063**
Video	1.352	1.312	1.301	**1.281**
DigiMu	1.363	1.313	1.171	**1.165**

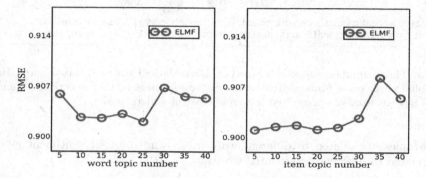

Fig. 3. Parameter tuning for the numbers of item- and word-topics. Smaller is better.

the model performance is relatively robust for both parameters, and a range between 5 and 20 usually give good performance.

4.3 Evaluation on Model Interpretability

Besides the performance, interpretability is also important to consider in rating prediction. Here, we conduct evaluation experiments to examine the model interpretability. We start with an illustrative example to show the explainable recommendation results for our model.

Qualitative analysis of the ELMF model. In specific, we are particularly interested in two queries: (1) what are the semantics for the subgroups of users and items, and (2) why a user likes or dislikes an item. To answer the first query, ELMF probabilistically clusters items using topic models, and each item-topic groups items with coherent semantics. We present three sample item-topics on *Music* dataset discovered by ELMF in Fig. 4. As we can see, the identified item-topics are clear and coherent. To answer the second query, we attach each topic with the most related words. Given an item-topic c, we select words by computing $\sum_{k=1}^{K_2} P(w|k)P(k|c,l)$ with a specific sentiment label l. Here, we consider two sentiment labels, *i.e.,* positive and negative. It is interesting to see different

Positive words: string, perfect, good, stand, solid, happy, capo, tight, standard, pick
Negative words: not-good, low, lack, amp, bet, feel, battery, band, wav, weight

Positive words: solid, pick, gig, happy, head, recommend, sturdy, tight, plenty
Negative words: money, no-fit, sign, start, bottom, hard, high, mic, disappoint, min

Positive words: mic, set, record, board, boss, switch, control, big, hear, elixir
Negative words: min, stiff, sign, design, preamp, headstock, clamp, hard, bar

Fig. 4. Three sample item-topics learned on *Music* dataset with associated words. Red and blue words come from positive and negative sentiment labels respectively. We use icons but not words to present each item for ease of understanding.

item-topics are related to different word-topics and different sentiment labels are related to different opinionated words.

Quality evaluation of the identified item-topics. Above, we have shown the sample item-topics learned by our model. Now we quantitatively evaluate the quality of these item-topics. We select BPMTMF as a baseline since it is also able to generate item-topics. Intuitively, a good item-topic should group items from same categories. Our datasets provide the original Amazon category labels of these items in a three-layer hierarchy. We select the *purity* as the evaluation metrics. *Purity*[1] is a popular measure to evaluate the clustering quality. It compares the generated clusters against the gold results. We take the categorization of items on Amazon as the ground truth for evaluation. We present the comparison results in Table 4. Since each item is associated with three category labels, we can compute three different purity scores. Overall, our method has better purity performance compared with BPMTMF. The major reason is that BPMTMF only utilizes rating information while our model further leverages review information.

Table 4. Purity comparison with three-level category labels.

Category level	1st	2nd	3rd
BPMTMF	0.659	0.622	0.246
ELMF	0.702	0.628	0.264

---

[1] http://nlp.stanford.edu/IR-book/html/htmledition/evaluation-of-clustering-1.html.

5   Conclusion

In this paper, we made an attempt to improve the interpretability of LMF methods by leveraging both item co-rated patterns and user reviews. We incorporated item-topics to construct meaningful subgroups, and associate them with opinionated word-topics to explain the semantics and sentiments for users ratings. By using topic-specific latent factors, our model yields competitive performance while the learned item- and word-topics give good interpretability to the recommendation results. Currently, our work only considers two sentiment labels, in the future, we will explore more levels of sentiments for more fine-grained explanations on the correlations between ratings and sentiments.

Acknowledgment. Xin Zhao was partially supported by the National Natural Science Foundation of China under grant 61502502 and the Beijing Natural Science Foundation under grant 4162032.

References

1. Ai, Q., Zhang, Y., Bi, K., Chen, X., Croft, W.B.: Learning a hierarchical embedding model for personalized product search. In: SIGIR. ACM (2017)
2. Amatriain, X., Mobasher, B.: The recommender problem revisited: morning tutorial. In: KDD (2014)
3. Blei, D.M., Ng, A.Y., Jordan, M.I.: Latent Dirichlet allocation. JMLR **3**(Jan), 993–1022 (2003)
4. Chen, C., Li, D., Zhao, Y., Lv, Q., Shang, L.: WEMAREC: accurate and scalable recommendation through weighted and ensemble matrix approximation. In: SIGIR. ACM (2015)
5. Chen, X., Qin, Z., Zhang, Y., Xu, T.: Learning to rank features for recommendation over multiple categories. In: SIGIR. ACM (2016)
6. Dhillon, I.S., Mallela, S., Modha, D.S.: Information-theoretic co-clustering. In: KDD. ACM (2003)
7. Ganu, G., Elhadad, N., Marian, A.: Beyond the stars: improving rating predictions using review text content. In: WebDB. Citeseer (2009)
8. He, X., Chen, T., Kan, M.Y., Chen, X.: Trirank: review-aware explainable recommendation by modeling aspects. In: CIKM. ACM (2015)
9. Koren, Y., Bell, R., Volinsky, C.: Matrix factorization techniques for recommender systems. Computer **42**(8) (2009)
10. Lee, J., Kim, S., Lebanon, G., Singer, Y.: Local low-rank matrix approximation. In: ICML (2013)
11. Xiao, L., Min, Z., Yongfeng, Z.: Joint factorization topic models for cross-city recommendation. In: Chen, L., Jensen, C.S., Shahabi, C., Yang, X., Lian, X. (eds.) APWeb-WAIM 2017. LNCS, vol. 10366, pp. 591–609. Springer, Cham (2017). doi:10.1007/978-3-319-63579-8_45
12. Liu, H., He, J., Wang, T., Song, W., Du, X.: Combining user preferences and user opinions for accurate recommendation. Electron. Commer. Res. Appl. **12**(1), 14–23 (2013)
13. Mackey, L.W., Jordan, M.I., Talwalkar, A.: Divide-and-conquer matrix factorization. In: NIPS (2011)
14. McAuley, J., Leskovec, J.: Hidden factors and hidden topics: understanding rating dimensions with review text. In: Recsys. ACM (2013)

15. McAuley, J., Pandey, R., Leskovec, J.: Inferring networks of substitutable and complementary products. In: KDD. ACM (2015)
16. Mnih, A., Salakhutdinov, R.: Probabilistic matrix factorization. In: NIPS (2008)
17. Salakhutdinov, R., Mnih, A.: Bayesian probabilistic matrix factorization using Markov Chain Monte Carlo. In: ICML. ACM (2008)
18. Tan, Y., Zhang, M., Liu, Y., Ma, S.: Rating-boosted latent topics: understanding users and items with ratings and reviews. In: IJCAI, pp. 2640–2646 (2016)
19. Wang, K., Zhao, W.X., Peng, H., Wang, X.: Bayesian probabilistic multi-topic matrix factorization for rating prediction. In: IJCAI (2016)
20. Wu, C.Y., Beutel, A., Ahmed, A., Smola, A.J.: Explaining reviews and ratings with PACO: Poisson additive co-clustering. In: WWW (2016)
21. Wu, Y., Ester, M.: FLAME: A probabilistic model combining aspect based opinion mining and collaborative filtering. In: WSDM. ACM (2015)
22. Zhang, Y.: Explainable recommendation: theory and applications. arXiv preprint arXiv:1708.06409 (2017)
23. Zhang, Y., Ai, Q., Chen, X., Croft, W.: Joint representation learning for top-n recommendation with heterogeneous information sources. In: CIKM. ACM (2017)
24. Zhang, Y., Lai, G., Zhang, M., Zhang, Y., Liu, Y., Ma, S.: Explicit factor models for explainable recommendation based on phrase-level sentiment analysis. In: SIGIR. ACM (2014)
25. Zhang, Y., Zhang, M., Liu, Y., Ma, S.: Improve collaborative filtering through bordered block diagonal form matrices. In: SIGIR. ACM (2013)
26. Zhang, Y., Zhang, M., Liu, Y., Ma, S., Feng, S.: Localized matrix factorization for recommendation based on matrix block diagonal forms. In: WWW, pp. 1511–1520. ACM (2013)

Finding and Quantifying Temporal-Aware Contradiction in Reviews

Ismail Badache[✉], Sébastien Fournier, and Adrian-Gabriel Chifu

LSIS UMR 7296 CNRS, University Aix-Marseille, Marseille, France
{ismail.badache,sebastien.fournier,adrian.chifu}@lsis.org

Abstract. Opinions (reviews) on web resources (e.g., courses, movies), generated by users, become increasingly exploited in text analysis tasks, the detection of contradictory opinions being one of them. This paper focuses on the quantification of sentiment-based contradictions around specific aspects in reviews. However, it is necessary to study the contradictions with respect to the temporal dimension of reviews (their sessions). In general, for web resources such as online courses (e.g. *coursera* or *edX*), reviews are often generated during the course sessions. Between sessions, users stop reviewing courses, and there are chances that courses will be updated. So, in order to avoid the confusion of contradictory reviews coming from two or more different sessions, the reviews related to a given resource should be firstly grouped according to their corresponding session. Secondly, aspects are identified according to the distributions of the emotional terms in the vicinity of the most frequent nouns in the reviews collection. Thirdly, the polarity of each review segment containing an aspect is estimated. Then, only resources containing these aspects with opposite polarities are considered. Finally, the contradiction intensity is estimated based on the joint dispersion of polarities and ratings of the reviews containing aspects. The experiments are conducted on the Massive Open Online Courses data set containing 2244 courses and their 73,873 reviews, collected from coursera.org. The results confirm the effectiveness of our approach to find and quantify contradiction intensity.

Keywords: Sentiment analysis · Aspect detection · Contradiction intensity

1 Introduction

Nowadays, web 2.0 has become a participatory platform where people can express their opinions by leaving traces (e.g., review, rating, like) on web resources. Many services, such as blogs and social networks, allow the generation of these traces. They represent a rich source of social data, which can be exploited in various contexts [1,2]. We mention in particular the field of sentiment analysis [9], where traces are exploited with the purpose of identifying a customer's attitude towards a product or its characteristics, or revealing the

© Springer International Publishing AG 2017
W.-K. Sung et al. (Eds.): AIRS 2017, LNCS 10648, pp. 167–180, 2017.
https://doi.org/10.1007/978-3-319-70145-5_13

reaction of people to an event. Such problems require rigorous analysis of the aspects covered by the sentiment to produce a representative and targeted result.

Another issue concerns opinion diversity on a given topic. Some work addresses it in different fields of research. For example, Wang and Cardie [24] aim to identify the sentiments of a sentence expressed during a discussion and using them as characteristics in a classifier that predicts dispute in the discussion. Socher et al. [19] automatically identify debates between users from textual content in forums, based on latent variable models. Other studies analyze user interactions, for example, extracting the *agreement* and *disagreement* expressions [15] and deducing the user relations by looking at their textual exchanges [8].

This paper investigates the entities (e.g. aspects) for which the contradictions can occur in the reviews associated with a web resource (e.g. movies, courses) and how to estimate their intensity. A contradiction can occur when there are conflicting opinions for a specific aspect, which is a form of sentiment diversity. Moreover, this contradiction can occur throughout a specific time period (session).

To design our approach, different fundamental tasks are aggregated: (1) clustering reviews according to their session; (2) identifying aspects characterizing these reviews; (3) analyzing sentiments of reviews to capture opposing opinions around each aspect; (4) using a measure of dispersion to estimate the intensity of these contradictory opinions. Furthermore, tests carried out on a set of real data (coursera.org), as well as a user study, demonstrate that our approach is able to identify effectively and significantly the contradictions and their intensity. The main contributions of this work can be summarized as it follows:

(C1). We present an approach for contradiction detection, which is based on sentiment-aspect extraction during each specific session of reviews. Therefore, we group reviews according to their sessions.

(C2). We formally estimate the contradiction intensity, and further describe two variations of the solution: *Averaged centroid* and *Weighted centroid*.

(C3). We experimentally evaluate the proposed approach by creating our own data set collected from coursera.org. In addition, we perform a user study.

The rest of this paper is structured as follows: Sect. 2 presents some related work and the background. Section 3 details our approach for detecting contradiction and compute its intensity. Section 4 reports on the results of our evaluation. Finally, Sect. 5 concludes this paper and announces our perspectives.

2   Background and Related Work

Contradiction detection is a complex process that requires the use of several state of the art methods (aspect detection, sentiment analysis). Moreover, to the best of our knowledge, very few studies treat the estimation of contradiction intensity. This section briefly presents the approaches related to aspect detection and sentiment analysis, which are useful for introducing our approach. Then, it presents some approaches of detecting controversies that are close to our work.

2.1   Aspect Detection Approaches

The first attempts to detect aspects were based on classical information extraction approaches using the frequent nominal sentences [10]. Such approaches work well for the detection of aspects that are in the form of single name, but are less useful when the aspects have low frequency. Similarly, other studies use Conditional Random Fields (CRFs) or Hidden Markov Models (HMMs) [6]. Other methods are unsupervised and have proven their effectiveness, such as [20], that built a Multi-Grain Topic Model, and [12] that proposed HASM (unsupervised Hierarchical Aspect Sentiment Model) which allow to discover a hierarchical structure of the sentiment, based on the aspects in the unlabelled online reviews. In our work, the explicit aspects are extracted using the unsupervised method presented in [17]. Poria et al., [17] proposed a rule-based approach that exploits common-sense knowledge and sentence dependency trees to detect both explicit and implicit aspects in product reviews.

2.2   Sentiment Analysis Approaches

Sentiment analysis has been the subject of much previous research. As in the case of the aspects detection, the supervised and the unsupervised approaches each propose their solutions. Thus, some unsupervised approaches are based on lexicons, such as the approach developed by [23] or corpus-based methods such as in [14]. Pang et al. [16] proposed supervised approaches, which perceive the task of sentiment analysis as a classification task and therefore use methods such as SVM (Support Vector Machines) or Bayesian networks. Other recent studies are based on the RNN (Recursive Neural Network), such as in [19]. In our work, sentiment analysis is only a part of the contradiction detection process, it is inspired by the work of [16] using a Bayesian classifier. Naive Bayes is a probabilistic model that gives good results in the classification of sentiments and generally takes less time for the training compared to models as SVM or RNN.

2.3   Contradiction and Controversy Detection Approaches

The most related works to ours include [3, 7, 21, 22]. They attempt to detect contradiction in text. There are 2 main approaches, where contradictions are defined as a form of textual inference and analyzed using linguistic technologies.

Harabagiu et al. [7] proposed an approach for contradiction analysis that exploits linguistic features (e.g., types of verbs), as well as semantic information, such as negation (explicit contradiction, e.g., "I love you - I do not love you") or antonymy (words that have opposite meanings, i.e., "hot-cold" or "light-dark"). Their work defined contradictions as textual entailment, when two sentences express mutually exclusive information on the same topic. Further improving the work in this direction, De Marneffe et al. [3] introduced a classification of contradictions consisting of seven types that are distinguished by the features that contribute to a contradiction, e.g., antonymy, negation, numeric mismatches which may be caused by erroneous data: "there are 7 wonders of the world - the number of wonders of the world are 9". They defined contradictions as a situation where "two sentences are extremely unlikely to be true when considered together". Tsytsarau et al. [21], [22] proposed a scalable solution for the contradiction detection problem. In their work, they studied the contradiction problem using sentiments analysis. The intuition of their contradiction approach is that when the aggregated value for sentiments (on a specific topic and time interval) is close to zero, while the sentiment diversity is high, the contradiction should be high. Another theme related to our work concern the detection of controversies and disputes. Among these studies, several treat the controversy on Wikipedia and particularly in the case of the comments that surround the modifications of Wikipedia pages [24]. Other studies try to detect controversies on specific domains, for example in news or in debate analysis [18]. Other studies try to be more generic and detect the controversy on web [11]. Ennals et al. [5] addressed the problem as a search of conflicting topics on the web through text patterns like "It is not correct that...". In addition, there is also a line of investigation known as controversy research. The aim is to identify whether Web contents deal with controversial topics (e.g. abortion, religion, same-sex marriage) and notify the user when the topic that they are searching is controversial [4].

Our work also has a certain proximity to previous efforts concerning the detection of contradiction in text. However, unlike previous works such as [3–5, 7], that defined contradiction based on linguistic features and numeric mismatches, our work defines contradiction as sentiment-based conflicting opinions for a specific aspects, which is a form of sentiment diversity. This kind of contradiction can occur at one specific point of time or throughout a certain time period. In addition, to the best of our knowledge, none of previous studies attempt to quantify the intensity of contradiction or of controversy. Our main goal is to measure contradiction intensity in reviews generated during specific period (session), by exploiting their ratings and polarities around the aspects.

3   Time-Aware Contradiction Intensity

To measure contradiction intensity during a session, two dimensions are jointly exploited: the polarity around the aspect as well as the rating associated with the review. We used a dispersion function, based on these dimensions, that esti-

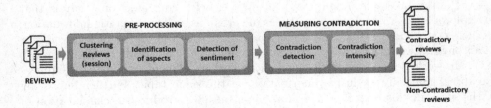

Fig. 1. Temporal sentiment-based contradiction intensity framework

mates the intensity between contradictory opinions (called after: reviews-aspect) (Fig. 1).

3.1 Pre-processing

The pre-processing module consists of 3 main steps: (1) clustering reviews according to their session; (2) aspects extraction from reviews; and (3) sentiment analysis of the text related to these aspects. We detail these steps in the following.

(a) Session-Based Clustering of Reviews. Generally, reviews are chronologically generated on resources, but some breaks (jumps) have been observed. These jumps (see Fig. 2) represent a silence of reviews generation by users during a specific period. Analyzing these jumps, we observed that the reviews are temporally related to the resource evolution, because this resource is often updated after each specific period. In order to properly handle the contradictions between the reviews, these reviews should be grouped according to their session. The sessions are defined for each resource according to specific X-days jump (silence) without reviews or without significant number of reviews (see Fig. 2).

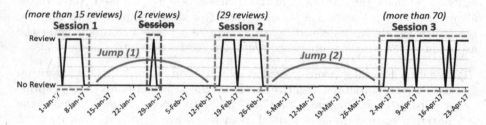

Fig. 2. Distribution of reviews in time of "Engagement & Nurture Marketing Strategies" course

To obtain these review groups, the following treatments are applied:

1. A threshold representing the jump duration is calculated for each course based on the mean of distances between reviews (e.g. the jump for the course "Engagement & Nurture Marketing Strategies" is: 35-days),
2. Grouping reviews according to the threshold for each course,
3. Identification of significant sessions by eliminating groups (false sessions) that contain insufficient number of reviews (no-dense sessions).

Remark. Only the groups (clusters) of reviews containing sufficient number of reviews are considered, i.e. for example, in Fig. 2 the group of reviews containing 2 reviews is ignored, hence, the using of K-Means [13]. In other words, k-means clustering is useful to partition the reviews into $k = 2$ sessions (dense or no-dense sessions (clusters) in terms of reviews quantity).

(b) **Extraction of Aspects.** In our study, an aspect is a frequently occurring nominal entity in reviews and it is surrounded by emotional terms. In order to extract the aspects from the reviews' text, we were inspired by the work of Poria et al., [17]. This method corresponds to our experimental data (*coursera* reviews). Additionally, the following treatments are applied:

1. Term frequency calculation of the reviews corpus,
2. Term categorization (part-of-speech tagging) of reviews using *Stanford Parser*[1],
3. Selection of terms having nominal category (NN, NNS)[2],
4. Selection of nouns with emotional terms in their five-neighborhoods (using *SentiWordNet*[3] dictionary),
5. Extraction of the most frequent (used) terms in the corpus among those selected in the previous step. These terms will be considered as aspects.

(c) **Sentiment Analysis.** The sentiments are represented by a real number in the range $[-1, 1]$ which indicates the polarity of the opinion expressed in the review segment with respect to an aspect (called review-aspect ra).

Pang's research [16] indicates that standard machine learning methods perform very well, even definitively outperforming human classifiers. Therefore, in order to estimate the sentiment of the review-aspect ra, we used Naive Bayes algorithm to predict sentiment polarity [16]. After several empirical experiments, the review-aspect ra is defined by an excerpt of 5 words before and after the aspect in review re. Our supervised sentiment model take into a account also:

(i) Negation handling (word preceded by *"no"*, *"not"*, *"n't"*). Our algorithm uses a state variable (Negative) to store the negation state. It transforms a word preceded by *"no"*, *"not"* or *"n't"* into *"not_"+word*. Whenever the negation state variable is verified, read words are treated as *"not_"+word*. The state variable is reset when a punctuation mark (*"?.,!:;"*) is encountered or when there is a double negation. The negative forms with respect to the normal forms of the same words are balanced during the training. This is to ensure that the number of *"not_"* forms is sufficient for the classification;
(ii) Combinations (*bigrams*) of adjectives with other words such as intensifiers and adverbs (e.g. *"very* bad" and *"absolutely* recommended").

3.2   Measuring Contradiction Intensity

Our application of time-aware contradiction analysis needs to follow the same steps that we previously identified for opinion mining, namely, session-based clustering of reviews, aspect identification and sentiment analysis.

A review on a given resource (e.g. courses, movies, media) and during a specific session covers one or more specific aspects (e.g. *lecturers* of courses, *actors*

---

[1] http://nlp.stanford.edu:8080/parser/.
[2] https://cs.nyu.edu/grishman/jet/guide/PennPOS.html.
[3] http://sentiwordnet.isti.cnr.it/.

of movies, etc.). For each review, some sentiments are expressed around these aspects. Then, we need to have a contradiction detection step, where individual sentiments (positive or negative) for each review-aspect ra_i are processed in order to reveal contradictory reviews-aspect ra_i. In this step, the goal is to efficiently combine the information extracted in the previous steps, in order to determine the aspects and time intervals (session) in which contradictions occur.

Definition. *Contradiction exists between two portions of review-aspect ra_1 and ra_2 containing an aspect with ra_1, $ra_2 \in D$ (Document), when the opinions (polarities) around the aspect are opposite (i.e. $pol(ra_1) \cap pol(ra_2) = \phi$).*

The main research problem addressed in this paper is related to the effective estimation of the contradictory opinions intensity in reviews related to specific aspects. The degree of contradiction around an aspect between the reviews is estimated using two dimensions: the polarity pol_i of the review-aspect ra_i and its rating rat_i. We assume that the greater the distance (i.e. dispersion) between these values related to each review-aspect ra_i of the same document D, the degree of contradiction is more important.

Let ra_i be a point on the plane with coordinates (pol_i, rat_i). The dispersion indicator with respect to the centroid $ra_{centroid}$ with coordinates $(\overline{pol}, \overline{rat})$ is defined as follows:

$$Disp(ra_{rat_i}^{pol_i}, D) = \frac{1}{n} \sum_{i=1}^{n} Distance(pol_i, rat_i) \tag{1}$$

$$Distance(pol_i, rat_i) = \sqrt{(pol_i - \overline{pol})^2 + (rat_i - \overline{rat})^2} \tag{2}$$

$Distance(pol_i, rat_i)$ represents the distance between the point ra_i of the scatter plot and the centroid $ra_{centroid}$ (see Fig. 3), and n is the number of ra_i. The two quantities pol_i and rat_i have different scale, it is essential to normalize them. The polarity pol_i is a probability, but the values of the ratings rat_i can be normalized as follows: $rat_i = \frac{rat_i - 3}{2}$ $(rat_i \in [-1, 1])$.

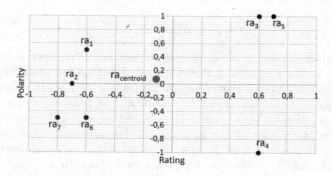

Fig. 3. Dispersion of reviews-aspect ra_i

By assigning each point ra_i having the same mass $1/n$, the indicator $Disp(ra_{rat_i}^{pol_i}, D)$ represents the divergence of the points ra_i with respect to the centroid $ra_{centroid}$.

- $Disp$ is positive or zero; $Disp = 0$ means that all ra_i are merged into $ra_{centroid}$ (no dispersion).
- $Disp$ increases when ra_i moved away from $ra_{centroid}$ (i.e. when the dispersion is increased).

The coordinates $(\overline{pol}, \overline{rat})$ of the centroid $ra_{centroid}$ can be calculated in two different ways. A simple way is to calculate the average of the points ra_i, in this case the centroid $ra_{centroid}$ corresponds to the average point of the coordinates $ra_i(pol_i, rat_i)$. Another finer way is to weight this average by the difference in absolute value between the two values of the coordinates (dimensions: pol_i, rat_i).

(i) Averaged Centroid i.e. centroid based on average of dimensions (polarity and rating). Let the statistical series with two variables (dimensions), where values are couples (pol_i, rat_i). The centroid (mean point of the series) based on the average of polarities and ratings is the point $ra_{centroid}$ in Fig. 3, which their coordinates are computed as follows:

$$\overline{pol} = \frac{pol_1 + pol_2 + ... + pol_n}{n}; \qquad \overline{rat} = \frac{rat_1 + rat_2 + ... + rat_n}{n} \qquad (3)$$

(ii) Weighted Centroid i.e. centroid based on the weighted average of dimensions. In this case, the coordinates of the centroid $ra_{centroid}$ are computed based on the weighted average of polarities and ratings as follows:

$$\overline{pol} = \frac{c_1 \cdot pol_1 + c_2 \cdot pol_2 + ... + c_n \cdot pol_n}{n}; \ \overline{rat} = \frac{c_1 \cdot rat_1 + c_2 \cdot rat_2 + ... + c_n \cdot rat_n}{n} \qquad (4)$$

where n is the number of points ra_i. The coefficient c_i is computed as follows:

$$c_i = \frac{|rat_i - pol_i|}{2n} \qquad (5)$$

In this two-dimensional vector representation, our hypothesis is that a point in this space is more important if the values of both dimensions are the most distant. We believe that a negative aspect in a review with a high rating has more weight and vice-versa. Consequently, a coefficient of importance for each point in space is calculated. This coefficient is based on the difference in absolute value between the values of the dimensions. The absolute value ensures that the coefficient is positive. The division by $2n$ represents a normalization by the maximum value of the difference in absolute value $(max(|rat_i - pol_i|) = 2)$ and n. For example, for a polarity of -1 and a rating of 1, the coefficient is $1/n$ $(|-1 - 1|/2n = 2/2n = 1/n)$, and for a polarity of 1 and a rating of 1, the coefficient is 0 $(|1 - 1|/2n = 0)$.

4   Experimental Evaluation

In order to validate our approach, a series of experiments was carried out on reviews collected from coursera.org. The objectives of these experiments are to:

1. evaluate the impact of sentiment-aspect on the detection of contradiction,
2. evaluate the impact of dispersion function based on polarity and rating to quantify contradiction *intensity*, using *averaged* and *weighted* centroid,
3. evaluate the effectiveness of contradiction detection based on reviews session.

4.1   Description of Test Data Set

To the best of our knowledge, no standard or annotated data set is available to evaluate the intensity of contradiction. Therefore, 2244 English courses are extracted from coursera.org via its API[4]. For each course, we have also collected its reviews, dates of reviews and ratings via the *parsing* of the course web pages (see the statistics in the Table 1).

Table 1. Statistics on coursera data set

Field	Total Number
Courses	2244
Courses Rated	1115
Reviews	73873
Reviews ★★★★★	1705
Reviews ★★★★★	1443
Reviews ★★★★★	3302
Reviews ★★★★★	12202
Reviews ★★★★★	55221

Table 2. List of detected aspects

Assignment	Content	Exercise
Information	Instructor	Knowledge
Lecture	Lecturer	Lesson
Material	Method	Presentation
Professor	Quality	Question
Quiz	Slide	Speaker
Student	Teacher	Topic
Video		
22 aspects		

Table 3. Statistics on some aspects extracted from the reviews of coursera.org

0.85 Aspects	#Rat 1	#Rat 2	#Rat 3	#Rat 4	#Rat 5	#Negative	#Positive	#Review	#Course
Content	176	179	341	676	1641	505	1496	1883	207
Lecture	185	206	290	613	1762	763	1508	1988	208
Video	228	238	356	707	1614	941	1421	2058	245

Table 3 presents some aspects among 22 useful aspects captured automatically from the reviews. To obtain judgments of contradictions and sentiments for a given aspect: (a) 3 assessors were asked to assess the sentiment class for each review-aspect; (b) 3 other assessors assessed the degree of contradiction between reviews-aspect. In average 6 reviews-aspect per course are judged manually for each aspect (totally: 1320 reviews-aspect of 220 courses i.e. 10 courses for each aspect). To evaluate sentiments and contradictions in the reviews-aspect of each

---

[4] https://building.coursera.org/app-platform/catalog.

course, 3-levels are used for sentiments: *Negative, Neutral, Positive*; and 5-levels for contradictions: *Not Contradictory, Very Low, Low, Strong* and *Very Strong* (Table 2).

We analyzed the agreement degree between assessors for each aspect using Kappa Cohen measure k. This indicator takes into account the proportion of agreement between the assessors and the proportion of agreement expected between the assessors by chance. The Kappa measure is equal to 1 if the assessors completely agree, 0 if they agree only by chance. k is negative if the agreement between assessors is worse than random. In our case, the k is 0.76 for sentiment assessors and k is 0.68 for contradiction assessors, which corresponds to a substantial agreement.

4.2   Results and Discussions

To evaluate the performance of our approach, correlation study was conducted (official measure on SemEval tasks[5]), by using the correlation coefficients of *Pearson* and *Spearman*, between the contradiction judgments given by the assessors and our obtained results.

Remarks: First, our sentiment analyzer takes as a training set 50,000 reviews of *IMDb* movies[6] (Due to the similarity of the vocabulary used in the reviews on

Table 4. Correlation results (WITHOUT Considering Reviews Session)

Measure	Config (1): averaged centroid	Config (2): weighted centroid
WITHOUT Considering Reviews Session		
(a) between contradiction judgments and approach results (sentiment accuracy: 79%)		
Spearman	0.42	0.49
Pearson	0.45	0.51
(b) between contradiction judgments and approach results (sentiment accuracy: 100%)		
Spearman	0.65	0.79
Pearson	0.68	0.87

Table 5. Correlation results (WITH Considering Reviews Session)

Measure	Config (1): averaged centroid	Config (2): weighted centroid
WITH Considering Reviews Session		
(a) between contradiction judgments and approach results (sentiment accuracy: 79%)		
Spearman	0.58*	0.69*
Pearson	0.61*	0.71*
(b) between contradiction judgments and approach results (sentiment accuracy: 100%)		
Spearman	0.70*	0.87*
Pearson	0.73*	0.91*

---

[5] http://alt.qcri.org/semeval2016/task7/.
[6] http://ai.stanford.edu/~amaas/data/sentiment/.

IMDb and *coursera*), and as a test set our reviews-aspect of *coursera*. Second, our sentiment analysis system provides an accuracy of 79% according to the correlation study. Third, assessors' judgments on sentiments are considered as perfect (reference) results and represent an accuracy of 100%.

Tables 4 and 5 summarize the correlation values obtained by taking into account the *averaged centroid* (Config (1)) and the *weighted centroid* (Config (2)) *WITH* and *WITHOUT* considering reviews session. In order to check the significance of the results (*WITH*) compared to (*WITHOUT*), we conducted the Student's t-test. The asterisk * is attached to the performance number of each row in Table 5 when *p-value < 0.05*. The results are discussed in the following.

(1) WITHOUT Considering Reviews Session

Config (1): averaged centroid. Table 4 show that the dispersion measurement based on the averaged centroid provides a positive correlation with judgments, Spearman: 0.42, 0.65 and Pearson: 0.45, 0.68, for the both cases: (a) 79% and (b) 100% sentiment accuracy, respectively. Indeed, the more polarities between the reviews-aspect are opposite, the more the set of reviews-aspect diverge from the centroid, hence the increased intensity dispersion. Moreover, the results obtained using the manual sentiments judgments (Table 4(b)) surpass those obtained using our sentiment analysis model (Table 4(a)) approximately with 50% (Spearman: 0.42 *Vs* 0.65 and Pearson: 0.45 *Vs* 0.68). Therefore, losing 21% in sentiments accuracy involves a 50% loss in detecting contradictions performance.

Config (2): weighted centroid. The results are also positive (Spearman: 0.49, 0.79 and Pearson: 0.51, 0.87). The results obtained by considering the coefficient c_i for each point of the space (review-aspect ra) are better compared to those obtained when this coefficient is ignored. These improvements are 16% (Spearman) using our sentiment model (Table 4(a)) and 22% (Spearman) using manual sentiment judgments (Table 4(b)). Indeed, the more divergent values of rating and polarity for every review-aspect, the higher is the impact on contradiction intensity. Also, the results of Config (2) presented in Table 4(b) are much better (Spearman: 0.79) than those presented in Table 4(a) (Spearman: 0.49). Therefore, the sentiment analysis model is an important factor that impacts the detection and the measurement of contradictions.

(2) WITH Considering Reviews Session

As previously, Table 5 shows a positive correlation for Config (1) and (2), with both assumptions in terms of sentiment accuracy (79% and 100%). The sentiment analysis model is always a factor that influences the results of the contradictions. Indeed, losing 21% in sentiments accuracy involves in average 23.5% loss in detecting contradictions performance. However, the results "WITH Considering Reviews Session" (see Table 5) show that the correlations values are better compared to those obtained when reviews session is ignored "WITHOUT Considering Reviews Session" (see Table 4). The comparative discussion is below.

Config (1): averaged centroid. The results (in Table 5(a) and (b)) show a significant improvement compared to those in Table 4(a) and (b). Indeed, when the *time* dimension (i.e. reviews are grouped by session) is taken into account, an improvement of 38% (Spearman) is recorded using our sentiment model in contradiction estimation (Table 5(a)) and 7% (Spearman) using manual sentiment judgments (Table 5(b)). From this comparison, we conclude that grouping reviews according to their session contributes to the effective contradiction detection. Moreover, the intensity of contradiction is estimated finely by taking into account only the reviews related to the specific session.

Config (2): weighted centroid. Using both the weighted centroid and the review session allows to improve the results even better than all previous runs. Compared to Config (2) in Table 4, the improvements are 49% (Spearman) using our sentiment model (Table 5(a)) and 10% (Spearman) using manual sentiment judgments (Table 5(b)). The difference comes from the advantage of the consideration of centroid based on the weighted average of dimensions (polarity and rating), as well as the clustering pre-processing of reviews (session).

Finally, we observe in all cases that our contradiction analysis approach, in terms of detection and intensity estimation, provides good results. The best results are obtained by Config (2) which takes into account the weighted centroid with considering temporal factor (session of reviews). According to t-test, the results show a statistically significant improvement. We believe that these improvements comes from the 3-steps pre-processing. Specifically, the grouping reviews according to their corresponding resources sessions, this contribute significantly to these well results. The dispersion formula measuring the intensity of contradiction becomes more effective when combined with an effective sentiment analysis model, which leads to a significant improvement of the results.

5   Conclusion

This paper introduced an approach that aims at estimating contradiction intensity, drawing attention to aspects in which users have contradictory opinions during a specific session. The intuition behind the proposed contradiction measure is that when the jointly dimensions (polarities and ratings) associated to reviews (on a specific aspect and session interval) are divergent (dispersed), while the sentiments diversity is high, then the contradiction should be high. Our study shows that contradiction exists if the sentiments around these reviews-aspect for the same resource are diverse. Clustering the reviews by sessions allow an effective treatment to avoid fake contradictions. Additionally, to quantify the contradiction, review-aspects are exploited using dispersion function, where more the coordinates polarities and ratings are opposite the more the impact is important on the contradiction intensity. The validation of our overall assumptions was examined on the data collection of coursera.org. The obtained results reveal the effectiveness of our approach. Finally, we note that we are aware that our approach of detecting contradiction is still limited. The major weakness of our approach is its dependence on the quality of sentiment analysis and aspect models. As the training set (IMDb reviews) is different from the test set (coursera reviews), if a word in the training set appears only in one class and does not appear in any other class, in this case, the classifier will always classify the text to that particular class. Moreover, the sentences are not processed, only predefined window of 5 words before and after the aspect is considered. Further scale-up experiments on other types of data are also envisaged. Even with these simple elements, the first results obtained encourage us to invest more in this track.

Acknowledgement. The project leading to this publication has received funding from Excellence Initiative of Aix-Marseille University - A*MIDEX, a French "Investissements d'Avenir" programme.

References

1. Badache, I., Boughanem, M.: Social priors to estimate relevance of a resource. In: IIiX, pp. 106–114 (2014)
2. Badache, I., Boughanem, M.: Fresh and diverse social signals: any impacts on search? In: CHIIR, pp. 155–164 (2017)
3. De Marneffe, M-C., Rafferty, A., Manning, C.: Finding contradictions in text. In: ACL, vol. 8, pp. 1039–1047 (2008)
4. Dori-Hacohen, S., Allan, J.: Automated controversy detection on the web. In: Hanbury, A., Kazai, G., Rauber, A., Fuhr, N. (eds.) ECIR 2015. LNCS, vol. 9022, pp. 423–434. Springer, Cham (2015). doi:10.1007/978-3-319-16354-3_46
5. Ennals, R., Byler, D., Agosta, J.M., Rosario, B.: What is disputed on the web? In: WICOW, pp. 67–74 (2010)
6. Hamdan, H., Bellot, P., Bechet, F.: Lsislif: Crf and logistic regression for opinion target extraction and sentiment polarity analysis. In: SemEval, pp. 753–758 (2015)
7. Harabagiu, S., Hickl, A., Lacatusu, F.: Negation, contrast and contradiction in text processing. In: AAAI, vol. 6, pp. 755–762 (2006)
8. Hassan, A., Abu-Jbara, A., Radev, D.: Detecting subgroups in online discussions by modeling positive and negative relations among participants. In: EMNLP (2012)
9. Htait, A., Fournier, S., Bellot, P.: Using web search engines for English and Arabic unsupervised sentiment intensity prediction. In: SemEval (2016)
10. Hu, M., Liu, B.: Mining and summarizing customer reviews. In: KDD (2004)
11. Jang, M., Allan, J.: Improving automated controversy detection on the web. In: SIGIR, pp. 865–868 (2016)
12. Kim, S., Zhang, J., Chen, Z., Oh, A., Liu, S.: A hierarchical aspect-sentiment model for online reviews. In: AAAI (2013)
13. MacQueen, J.: Some methods for classification and analysis of multivariate observations. In: Berkeley Symposium on Mathematical Statistics and Probability (1967)
14. Mohammad, S.M., Kiritchenko, S., Zhu, X.: Nrc-canada: Building the state-of-the-art in sentiment analysis of tweets. In: SemEval (2013)
15. Mukherjee, A., Liu, B.: Mining contentions from discussions and debates. In: KDD, pp. 841–849 (2012)
16. Pang, B., Lee, L., Vaithyanathan, S.: Thumbs up?: sentiment classification using machine learning techniques. In: EMNLP, pp. 79–86 (2002)
17. Poria, S., Cambria, E., Ku, L., Gui, C., Gelbukh, A.: A rule-based approach to aspect extraction from product reviews. In: SocialNLP (2014)
18. Qiu, M., Yang, L., Jiang, J.: Modeling interaction features for debate side clustering. In: CIKM, pp. 873–878 (2013)
19. Socher, R., Perelygin, A., Wu, J.Y., Chuang, J., Manning, C.D., Ng, A.Y., Potts, C.: Recursive deep models for semantic compositionality over a sentiment treebank. In: EMNLP, vol. 1631, p. 1642 (2013)

20. Titov, I., McDonald, R.: Modeling online reviews with multi-grain topic models. In: WWW, pp. 111–120 (2008)
21. Tsytsarau, M., Palpanas, T., Denecke, K.: Scalable discovery of contradictions on the web. In: WWW, pp. 1195–1196. ACM (2010)
22. Tsytsarau, M., Palpanas, T., Denecke, K.: Scalable detection of sentiment-based contradictions. DiversiWeb, WWW (2011)
23. Turney, P.D.: Thumbs up or thumbs down?: semantic orientation applied to unsupervised classification of reviews. In: ACL, pp. 417–424 (2002)
24. Wang, L., Cardie, C.: A piece of my mind: a sentiment analysis approach for online dispute detection. In: ACL, pp. 693–699 (2014)

Automatic Query Generation from Legal Texts for Case Law Retrieval

Daniel Locke[(✉)], Guido Zuccon, and Harrisen Scells

Queensland University of Technology, Brisbane, Australia
{daniel.locke,harrisen.scells}@hdr.qut.edu.au, g.zuccon@qut.edu.au

Abstract. This paper investigates automatic query generation from legal decisions, along with contributing a test collection for the evaluation of case law retrieval. For a sentence or paragraph within a legal decision that cites another decision, queries were automatically generated from a proportion of the terms in that sentence or paragraph. Manually generated queries were also created as a ground to empirically compare automatic methods. Automatically generated queries were found to be more effective than the average Boolean queries from experts. However, the best keyword and Boolean queries from experts significantly outperformed automatic queries.

1 Introduction

In common law jurisdictions, such as the United Kingdom, United States, and Australia, where the doctrine of precedent (*stare decisis*) is applicable, finding relevant and therefore binding legal principles is crucially important so that lawyers can discharge their duties to the court.[1]

Finding factually or legally applicable case law forms a large part of a lawyer's work. Studies have found that lawyers spend roughly 15 h per week finding case law, or as much as 28% of their yearly working hours [12,14]. With research playing a fundamental role in a lawyer's work, increasing the quality of legal research tools is of great importance. Despite this, little research has been conducted as to retrieval of case law, previous research has typically focused on either argumentation retrieval, ontological frameworks and case-based reasoning or retrieval in a discovery setting.[2]

This paper explores the use of automatic methods for the generation of queries for case law retrieval. These methods could be integrated: (i) in a contextual suggestion system that provides lawyers with relevant cases as they write; or (ii) within search functionalities, to support lawyers in the formulation of effective queries.

---

[1] The doctrine of precedent requires, broadly speaking, that like circumstances are considered in a like fashion; a case that considers a certain set of factual circumstances therefore must be followed for any future circumstances that are analogous.

[2] The obligation of parties to litigation to disclose all documents relevant to issues between them.

© Springer International Publishing AG 2017
W.-K. Sung et al. (Eds.): AIRS 2017, LNCS 10648, pp. 181–193, 2017.
https://doi.org/10.1007/978-3-319-70145-5_14

As automated methods, we evaluated a number of common keyword extraction methods, namely proportional inverse document frequency (IDF-r) [10], Kullback-Leibler divergence for informativeness (KLI) [17], and parsimonious language models (PLM) [6]. These methods aim to select appropriate terms as candidate queries from portions of legal documents.

We compared the investigated automatic methods to the original sentences and paragraphs that reference a previous legal decision. As an additional comparison, we also collected a number of queries (Boolean and best-match based) manually built by a legal expert. A new test collection was created to empirically evaluate the methods and investigation, and relevance assessments were made by a legal expert. This collection, which comprises of 63,916 documents, 100 topics, 248 manually created queries and a total of 2645 relevance assessments, is an additional contribution of this paper to the research in case law retrieval.

The paper continues as follows. In Sect. 2 we describe related work, including highlighting the lack of an adequate test collection for evaluating retrieval and query generation methods for case law retrieval. In Sect. 3 we describe the query generation methods investigated in this work, along with details on the creation of the test collection. In Sect. 4 we describe the experimental settings and the results of the empirical evaluation. Section 5 concludes this paper and provides an account of future work.

2   Related Work

Legal Information Retrieval. Early work in case law retrieval has focused on inference networks and comparing natural language and Boolean queries for retrieval of case law [18]. More broadly, and more recently, legal information retrieval (IR) has focused on: (1) question answering (Q&A) (the ResPubliQA task, which considered Q&A tasks involving approximately 10,000 legislative texts of the European Union [13], and the COLIEEE collection, which involved Q&A over a smaller number of Articles of the Japanese Civil Code [7], are both recent examples of this); (2) discovery, with TREC conducting from 2006, the Legal Track [2]; (3) argument retrieval (for instance see the work of Grabmair et al. [5]); and (4) network analysis of citations (for instance see van Opijnen [20]).

Lack of a test collection for Legal IR. The most notable gap in the area of Legal IR is the lack of an existing standardised test collection. While legal decisions are now more easily accessible and more freely available, no standard corpus exists for testing information retrieval of case law. This can, perhaps, be contrasted with the test collections used in the TREC legal discovery tracks.

Only two collections that contain case law have been created for legal IR [8,9], both of which were used for analysis of diversification. Koniaris et al. [8], who also acknowledged the lack of "standard testing data", created the largest collection, which contains 63,000 decisions of the United States Supreme Court, 300 queries and automatically generated relevance judgments (see below for details). This collection takes as its queries, a subset of the areas of law found in the Westlaw

Digest.[3] These queries, however, are unsuitable for the task investigated in this paper. In Table 1, we have compared queries used by Koniaris et al. [8] with those of Turtle [18], which represent queries created by an expert searcher (a lawyer), rather than extracted from the Westlaw Digest; this shows the artificial nature and the broad scope of the queries from Koniaris et al. Note that the collection compiled by Turtle is not publicly available. The table also shows for comparison a query from our collection that was created by a domain expert.

Table 1. Queries previously employed in legal information retrieval studies and an example of our queries. Koniaris et al.'s [8] queries were artificially created from Westlaw Digest topics. Turtle's [18] queries were created by lawyers.

Source	Query	Generation method
Koniaris et al. [8]	Products liability	Topic in Westlaw Laws of America Digest
Turtle [18]	(741 +3 824) FACTOR ELEMENT STATUS FACT /P VESSEL SHIP BOAT /p (46 +3 688) "JONES ACT" /P INJUR! /S SEAMAN CREWMAN WORKER	Manually created by expert searcher
This work	"sovereign immunity" AND (immunity OR indemnif!) AND state AND suit AND (surrend! OR exist!) AND (tribe OR tribal OR "indian trib!")	Manually created by expert searcher

Furthermore, the method used by Koniaris et al. [8] to generate relevance judgments does not lend itself to the task investigated in this paper. In that work, relevance judgments were created automatically from an LDA topic model created over the top-n results of each query, after which an acceptance threshold of 20% was taken to obtain a relevance decision for a given query. An alternative, realistic, determination of query relevance would be preferable: in this paper we use a domain expert to provide relevance assessments.

Queries in legal IR. Schweighofer and Geist [16] note that, unlike in other tasks, the effectiveness of Boolean queries in the legal domain might not be inferior to that of queries in best match retrieval. This is because, they argue, lawyers are domain experts and will necessarily have knowledge of synonyms, without which effectiveness may suffer. However, as they note, domain knowledge has its limits, and one cannot reasonably know all other possible choices for a word. Turtle makes similar comments [18,19], suggesting that the larger the collection searched on, the greater the difficulty in creating an effective Boolean

---

[3] A keynumber system of categorised areas and subareas of law. Areas of law can be searched or browsed by number.

query. Despite this, Poje [14] found that lawyers in practice for 2 years or less favoured natural language queries, whereas lawyers with more than 2 years practice favoured Boolean queries. For this reason, the collection that we contribute contains both Boolean and keyword queries.

Keyword extraction and query reduction. Keyword extraction (or key terms selection) consists of identifying appropriate terms "capable of representing information content" [15]. The goal of a keyword extraction method is to extract and rank keywords from an information object (a sentence, document or collection) [21]. Verberne et al. have investigated six unsupervised keyword extraction methods across a number of tasks, including from news retrieved for Boolean queries and for which keywords were extracted for the purpose of query suggestion. Keyword extraction methods are relevant to our work because they can be used to select appropriate keywords to form queries that retrieve relevant case law. In legal IR, the task of keyword extraction has been generally referred to as catchphrase or catchword identification; automatic methods for this task have been shown effective for legal document summarisation [4].

Query reduction consists of identifying one or more subsets of an original verbose query that allows for the better retrieval of relevant information. Query reduction is akin, in practice, to keyword extraction in that verbose queries that are to be reduced can be used in place of information objects as inputs of keyword extraction methods. Kumaran and Carvalho's [11] methods rely on the generation of shorter subqueries from an initial query, for which a classifier is used to predict the quality of a given subquery. Bendersky and Croft [3] developed an unsupervised method for extracting key concepts from verbose queries. Both the methods of Kumaran and Carvalho [11] and Bendersky and Croft [3] rely on the generation of permutations of sub-queries. This renders the methods not feasible for (very) long text, like the sentences and paragraphs used in our work (generating all possible sub-queries for text of n terms requires $n!$ combinations). Our work shares some similarities with the recent study by Koopman et al. [10], who investigated the generation of clinical queries from medical narratives. The similarities are that, like us, they also used proportional IDF (IDF-r) for query term selection, and they also studied the automatically generated queries with respect to queries issued by domain experts.

No prior work has examined the application of keyword extraction or query reduction methods for the automatic generation of queries for case law retrieval.

3   Methods

3.1   Creation of Test Collection

The collection contained 63,916 decisions (cases) of the United States Supreme Court (USSC)[4]. For each document, we included the title, the plain text, the HTML, the date the decision was filed, and a list of cited opinions. The HTML

---

[4] Decisions were downloaded from http://courtlistener.com.

was the whole of the text of a document, generally scanned from a pdf. We created a plain text representation by removing html tags. The average document length (for the plain text field) was 1,918 words.

To create topics for evaluation, we selected 50 cases from the collection, among the most recent in the collection, so that each case referred to two or more topically separate decisions. From these, we manually selected citations of 100 decisions, and built a corresponding set of 100 topics. For each topic, we included the sentence from the original case that cited the decision to be found, and the paragraph in which the sentence was contained. A topic was a sentence in a case (decision) of the United States Supreme Court. A case and therefore 2 topics were included if the 2 decisions that became separate topics:

1. were cited in the opinion of the court rather than the syllabus (court added headnote summary);
2. did not quote the cited case for a significant portion of the sentence;
3. cited a case that was within the collection (i.e. from the USSC);
4. were not for the same case;
5. were not two citations for the same proposition on separate occasions;
6. were not a citation where the court says it granted *"certiorari"*[5];
7. were not in the arguments advanced by a party to the case, unless it was an argument that became stated as a proposition of law;
8. were, if contained in a sentence that cited more than one case for one proposition, included as one citation instance with all other citations for the propostion, provided they met the above criteria;
9. were, if a sentence contained multiple citations for separate clauses, i.e. for different legal propositions, included as separate topics with the separate clauses as the cited case.

We describe the process to generate queries from the topics in the next subsections.

To select the evaluation measures to be used in the empirical experiments we further considered the task at hand. The query generation methods studied here are intended to be used to automatically retrieve relevant cases when lawyers write. In this case, the methods would retrieve relevant prior cases, which would be served as contextual recommendations to the user. These methods can also be used to help lawyers construct queries or suggest queries when exploring a collection of prior legal cases and decisions. This, and the availability of citation indices through which cited decisions can be traced, means in our view, in both scenarios users would only be interested in and inspect a handful of search results/suggestions. Thus, we identify precision at rank 1 and 5 (P@1 and P@5) as suitable evaluation measures for this task. We also consider average precision at rank 5 (AP@5) so as to attribute more importance to ranking relevant decisions early on. In addition, we calculate mean reciprocal rank (MRR), as it indicates, on average, at what rank the first relevant decision is identified. Note

---

[5] A statement by the Court as to whether it would grant review of a lower court's decision.

that given the way we built our collection, there is always at least one relevant decision given a case (the decision that is cited).

Relevance assessments were created by pooling the manually generated queries, the baseline sentences and paragraphs (used in a standard BM25 system, see Sect. 4.1) and the automatic query generation methods investigated in this work. Pooling was performed to guarantee that runs had nearly complete assessments for the target precision evaluation measures. Assessments were provided by the first author of this paper (a lawyer, but not a legal practitioner) using a purposely created web interface. A total of 2,593 assessments were made.

Relevance was determined on the basis of the extracted sentence for the topic. If the decision being assessed was the cited decision, it was determined to be relevant automatically. If the decision cited the same case as the sentence, and it was for the same or a legally similar proposition it was relevant. Otherwise the first couple of paragraphs of the decision were read to determine the issue of the case, and a keyword search of the decision was made to determine relevance. If the decision was on the topic broadly, being that it would be useful for a lawyer to read or at least skim, then it was classified as relevant. We did not take the full list of cited decisions available for a case as relevant as a case may cite a number of different decisions for any number of different topics.

3.2   Manual Query Generation

As a baseline, for each topic, we evaluated both the topic sentence, and the paragraph that contained the topic sentence, with any of the cited case name removed. On average, sentences contained 47.17 terms, while paragraphs contained 148.13 terms. Sentences and paragraphs were also the information objects given as input to the automatic query generation methods.

For each topic, we also created one to three Boolean queries. These queries were created by the first author of this paper. Boolean queries were used because lawyers commonly use Boolean queries to search for case law in the existing legal systems (see Poje [14], and Sect. 2). A total of 248 Boolean queries were manually created (on average, 2.48 queries per topic) by identifying important keywords from each citation (topic) and from introducing keywords relevant to an area of law based on the first author's domain knowledge.

Boolean queries were also transcribed by removing the Boolean operators to be used in best-match retrieval systems (a match query in Elasticsearch).

Finally, keyword-based queries were created by the legal expert by manually listing keywords that were relevant to the topic or area of law, from a skim read of the issues in the case or from the paragraph or sentence, and included other words where a topic might usually use such words as synonyms. This was done at the same time as the creation of the Boolean queries.

3.3   Automatic Query Generation

For both topic sentences and paragraphs (information objects), we investigated the following automatic query generation methods: (i) the IDF-r method [10],

(ii) the Kullback-Liebler informativeness (KLI) [17], and (iii) the parsimonious language model (PLM) [6].

For all methods, we considered the list of terms T contained in an information object (a sentence or paragraph). The methods are used to produce a subset list T'. This list is produced by ranking the terms according to the specific method used (e.g., by IDF score for the IDF-r method), where scores indicate how relevant a term is for describing the information object, then selecting the top n terms from this ranking, according to a rank cut-off (or proportion parameter). For example, IDF-r selects the $\lceil \frac{|T|}{r} \rceil$ terms in T with the highest IDF score.

For the KLI method, we used statistics for terms in our collection to compute the probability of a term t in an information object D (sentence or paragraph in our case), $P(t|D)$. Conversely, to compute $P(t|C)$ we used statistics for terms in a background language model, computed using a general purpose collection. With these statistics we were able to compute the KLI score for a term; formally:

$$KLI(t) = P(t|D)log\frac{P(t|D)}{P(t|C)} \tag{1}$$

For the PLM method, we used our collection as the background language model, $P(t|C)$, and the information object, D (the sentence or paragraph) as the foreground language model. Probabilities were estimated using the expectation maximization algorithm, with the steps defined as:

$$E-step: \quad e_t = tf(t,D)\frac{\lambda P(t|D)}{(1-\lambda)P(t|C) + \lambda P(t|D)} \tag{2}$$

$$M-step: \quad P(t|D) = \frac{e_t}{\sum_{t' \in D} e_{t'}} \tag{3}$$

where $\lambda \in [0,1]$ is a smoothing parameter that controls the influence of statistics from the collection (C) over the statistics from the information object (D).

4   Experiments

4.1   Experimental Settings

We indexed the collection (plain text part) using Elasticsearch version 5.4.2.[6] When indexing, we used Porter stemming and no stop words. As the retrieval function we used the default in Elasticsearch (BM25 with $b = 0.75$, $k1 = 1.2$). We did not tune these parameters as no ground truth was available for tuning at retrieval time (relevance assessments were performed once runs were pooled).

Queries were processed also with Elasticsearch and using the Porter stemmer; stop words were removed from queries. For the Boolean queries, these were formatted using the Elasticsearch query syntax for Boolean queries; all other queries were treated as *match* queries in Elasticsearch (i.e. standard best match).

---

[6] https://www.elastic.co/.

Sentences (s) and paragraphs (p) were used as information objects from which to generate queries with the automatic methods studied in this paper. Before using the automatic methods, information objects were stripped of stop words. Results were analysed with respect to these different information sources.

For PLM, we explored settings of the smoothing parameter λ[7] between 0.1 (PLM is dominated by the statistics of the collection language model) to 1 (no smoothing: all statistics are computed from the topic sentence or paragraph), with step 0.1. These results are analysed separately (see below), and only the best settings for PLM are compared with the other methods. The background collection was Clueweb12B, while our collection was used as the foreground.

All results were analysed for different levels of the proportion parameter r, which dictates how many of the original terms (from the sentences or the paragraphs) were required to be selected to generate the queries. For example, $r = 0.1$ indicates that 10% of the total number of terms in the information objects were selected (rounded to the upwards integer). Note, $r = 0$ corresponds to selecting one term only.

4.2   Results for Manual Queries

Table 2 reports the retrieval effectiveness for the manually generated queries, along with the sentence (S) and paragraph (P) baselines. For the Boolean queries, we report both the mean average effectiveness achieved by the different query variations of the topics (Bavg – recall that, for every topic, between 1 and 3 Boolean variations were obtained) and the mean of the best effectiveness achieved by Boolean queries for each topic (Bbest). The corresponding values are also reported for the Boolean queries from which we removed the Boolean operands (NBavg and NBbest). Finally, we also report the effectiveness achieved when manually selecting keyword queries from the sentences (K).

The results highlight that Boolean queries are largely outperformed by non Boolean queries; specifically NBbest outperforms all other manual queries, with K also performing above all methods, but NBbest. Interestingly, the Bavg and NBavg are outperformed by querying using the whole of a topic sentence or paragraph. With respect to Bavg and Bbest, their performance may be hindered as a result of some queries being too specific or restrictive: a total of 36 queries returned no results, and a total of 9 topics returned no results. Further, this, in combination with the performance of Bbest and NBbest compared to Bavg and NBavg shows that manually created queries do not perform consistently: the same expert user formulated queries that largely varied in effectiveness. This finding is in line with previous research on query variations in other domains [1,22].

4.3   Tuning of PLM

The Parsimonious language model method is characterised by a smoothing parameter that controls the influence of statistics from the collection over the statis-

---

[7] λ is responsible for smoothing between the background language model (the legal collection), and the foreground language model (the sentence or paragraph).

Table 2. Effectiveness of manual queries and baselines on the 100 topics in the collection. For Boolean queries, multiple queries were devised for each topic: Bavg and NBavg refer to the mean average effectiveness over all query variations for all topics (with Boolean queries, and with the keywords in the Boolean queries but no Boolean operators); Bbest and NBbest refer to the average effectiveness for the best query of each topic. Statistical significant differences (paired t-test, $p < 0.05$) compared to the baselines are reported with * (for topic sentences, S) and † (for topic paragraphs, P).

	P@1	P@5	AP@5	MRR
Bavg	0.4400*†	0.3140*†	0.1528*†	0.4844*†
Bbest	0.7012	0.6500	0.2530	0.4840
NBavg	0.5733*†	0.3913*	0.2393*†	0.6362*†
NBbest	0.8377*	0.7800*†	0.5520*	0.3441*
K	0.7300	0.5020†	0.3121	0.8033
S	0.6800	0.4540	0.2958	0.7520
P	0.6800	0.4440	0.3015	0.7656

tics from the information object (the topic sentence or paragraph). We studied the impact of this parameter on effectiveness (AP@5). Figure 1 reports AP@5 for varying levels of λ and the proportional cutoff value r.

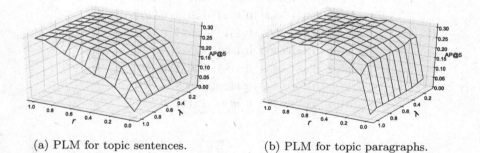

(a) PLM for topic sentences.   (b) PLM for topic paragraphs.

Fig. 1. AP@5 for PLM for varying values of the smoothing parameter λ and the proportion cutoff r, applied to topic sentences (left) and paragraphs (right).

While results are heavily affected by the proportion of terms selected when generating a query (r), there appear to be no differences in effectiveness due to different settings of the smoothing parameter λ, when $\lambda \leq 0.8$ and sentences are considered as input. The situation is similar when applying PLM on paragraphs. This may be due to the fact that smoothing only affects a limited number of not relevant documents. Because of these results, we fix $\lambda = 0.5$ when comparing PLM to the other query generation methods.

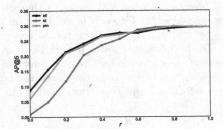 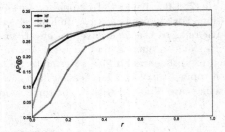

(a) Effectiveness of automatic generation methods for topic sentences. (b) Effectiveness of automatic generation methods for topic paragraphs.

Fig. 2. AP@5 for the automatic generation methods for varying values of the proportion cutoff r, when applied to topic sentences (left) and paragraphs (right).

Table 3. Effectiveness of automatically generated queries using sentences (s) and paragraphs (p) as input information objects, as well of baselines. The methods are reported using the configuration of r (proportion of query terms) that provided the highest effectiveness in terms of AP@5. Statistical significant differences (paired t-test, $p < 0.05$) compared to the baselines are reported with $\dagger$ (for topic paragraphs, P).

	r	P@1	P@5	AP@5	MRR
S	-	0.6800	0.4540	0.2958	0.7520
P	-	0.6800	0.4440	0.3015	0.7656
idf(s)	0.9	0.6800	0.4540	0.2958	0.7522
kli(s)	0.8	0.6900	0.4540	0.2974	0.7591
plm(s)	0.8	0.6900	0.4560	0.2987	0.7593
idf(p)	0.6	0.6800	0.4480	0.3053	0.7556
kli(p)	0.7	0.7000	0.4420	0.3068	0.7671
plm(p)	0.6	0.6900	0.4640	0.3114	$0.7629^{\dagger}$

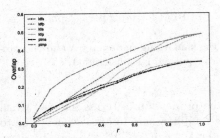 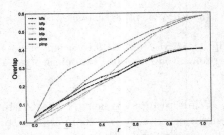

(a) Overlap between automatic generation methods and Bbest. (b) Overlap between automatic generation methods and K.

Fig. 3. Number of query terms in common (overlap) between Boolean queries and keywords, and automatically generated queries, for increasing levels of r.

4.4   Results for Automatically Generated Queries

We now analyse the effectiveness of automatic query generation methods with respect to r, the number of query terms: this is reported in Fig. 2. Methods plm(p) (PLM on paragraphs) and plm(s) (on sentences) both outperform baselines S and P (see Table 3) and display the best performance for automatic reduction methods, except kli(p) for P@1 and MRR when the proportion that takes the maximum AP@5 is found. Methods plm(p), plm(s) and idf(p) achieve the highest results at earlier proportions compared to kli(p) and the other methods of reduction from S. The effectiveness of the PLM methods are also evidenced through the higher percentage of overlap with the manually created queries, as shown in Fig. 3. While this is the case, all methods of reduction from S only achieve their best results at very high proportions (nearing the whole of the sentence), and over half of the paragraph for reduction methods from P. Nevertheless, these results are sensibly lower than some of the manual queries devised by the legal expert, e.g., Bbest and NBbest. Perhaps therefore, methods to introduce new terms into a query (i.e. query expansion), rather than reduce a query to distill terms are an appropriate area of investigation.

5   Conclusion

In this work we have considered automatic query generation methods for case law retrieval, and compared these methods with manual queries akin to those prepared by legal experts. We found that existing keyword extraction methods are as effective as average Boolean queries issued by experts. However, we also found that automatic methods are substantially inferior to keyword queries, and the best Boolean queries from experts. These effective queries often use terms not mentioned in the cases for which the queries are designed. The query generation methods we considered only select terms from portions of the cases at hand, thus not inferring additional relevant terms not mentioned in the cases. Methods that introduce new terms into a query (i.e. query expansion) are an appropriate area for future investigation.

As part of this research, we have also contributed a test collection for case law retrieval. While our collection contains a number of manual relevance assessments that allows us to reliably evaluate the methods considered here, more assessments are required to evaluate other methods. Future work will extend this collection to make it reusable for evaluation for other case law retrieval methods. Our collection, retrieval runs and analysis are available online at https://github.com/ielab/ussc-caselaw-collection.

References

1. Bailey, P., Moffat, A., Scholer, F., Thomas, P.: User variability and ir system evaluation. In: Proceedings of the 38th International ACM SIGIR Conference on Research and Development in Information Retrieval, pp. 625–634. ACM (2015)

2. Baron, J.R., Lewis, D.D., Oard, D.W.: Trec 2006 legal track overview. In: The Fifteenth Text REtrieval Conference (TREC 2006) Proceedings (2006)
3. Bendersky, M., Croft, W.B.: Discovering key concepts in verbose queries. In: Proceedings of the 31st Annual International ACM SIGIR Conference on Research and Development in Information Retrieval, pp. 491–498. ACM (2008)
4. Galgani, F., Compton, P., Hoffmann, A.: Combining different summarization techniques for legal text. In: Proceedings of the Workshop on Innovative Hybrid Approaches to the Processing of Textual Data, pp. 115–123. Association for Computational Linguistics (2012)
5. Grabmair, M., Ashley, K.D., Chen, R., Sureshkumar, P., Wang, C., Nyberg, E., Walker, V.R.: Introducing luima: an experiment in legal conceptual retrieval of vaccine injury decisions using a uima type system and tools. In: Proceedings of the 15th International Conference on Artificial Intelligence and Law, ICAIL 2015, pp. 69–78 (2015)
6. Hiemstra, D., Robertson, S., Zaragoza, H.: Parsimonious language models for information retrieval. In: Proceedings of the 27th Annual International ACM SIGIR Conference on Research and Development in Information Retrieval, pp. 178–185. ACM (2004)
7. Kim, M.-Y., Xu, Y., Lu, Y., Goebel, R.: Legal question answering using paraphrasing and entailment analysis. In: Tenth International Workshop on Juris-Informatics (JURISIN) (2016)
8. Koniaris, M., Anagnostopoulos, I., Vassiliou, Y.: Multi-dimension diversification in legal information retrieval. In: Cellary, W., Mokbel, M.F., Wang, J., Wang, H., Zhou, R., Zhang, Y. (eds.) WISE 2016. LNCS, vol. 10041, pp. 174–189. Springer, Cham (2016). doi:10.1007/978-3-319-48740-3_12
9. Koniaris, M., Anagnostopoulos, I., Vassiliou, Y.: Evaluation of diversification techniques for legal information retrieval. Algorithms **10**(1), 22 (2017)
10. Koopman, B., Cripwell, L., Zuccon, G.: Generating clinical queries from patient narratives. In: Proceedings of the 40th International ACM SIGIR Conference on Research and Development in Information Retrieval (to appear, 2017)
11. Kumaran, G., Carvalho, V.R.: Reducing long queries using query quality predictors. In: Proceedings of the 32nd International ACM SIGIR Conference on Research and Development in Information Retrieval, pp. 564–571. ACM (2009)
12. Lastres, S.A.: Rebooting legal research in a digital age. Technical report, LexisNexis (2013)
13. Peñas, A., et al.: Overview of ResPubliQA 2009: question answering evaluation over european legislation. In: Peters, C., Di Nunzio, G.M., Kurimo, M., Mandl, T., Mostefa, D., Peñas, A., Roda, G. (eds.) CLEF 2009. LNCS, vol. 6241, pp. 174–196. Springer, Heidelberg (2010). doi:10.1007/978-3-642-15754-7_21
14. Poje, J.: Legal research. American Bar Association Techreport 2014 (2014)
15. Salton, G.: Automatic Information Organization and Retrieval. McGraw Hill Text, New York (1968)
16. Schweighofer, E., Geist, A.: Legal query expansion using ontologies and relevance feedback. In: CEUR Workshop Proceedings, vol. 321, pp. 149–160 (2007)
17. Tomokiyo, T., Hurst, M.: A language model approach to keyphrase extraction. In: Proceedings of the ACL 2003 Workshop on Multiword Expressions: Analysis, Acquisition and Treatment, vol. 18, pp. 33–40. Association for Computational Linguistics (2003)
18. Turtle, H.: Natural language vs. boolean query evaluation: a comparison of retrieval performance. In: Proceedings of the 17th Annual International ACM SIGIR Conference on Research and Development in Information Retrieval, pp. 212–220 (1994)

19. Turtle, H.: Text retrieval in the legal world. Artif. Intell. Law **3**(1), 5–54 (1995)
20. van Opijnen, M.: Citation analysis and beyond: in search of indicators measuring case law importance. In: JURIX, vol. 250, pp. 95–104 (2012)
21. Verberne, S., Sappelli, M., Hiemstra, D., Kraaij, W.: Evaluation and analysis of term scoring methods for term extraction. Inform. Retrieval J. **19**(5), 510–545 (2016)
22. Zuccon, G., Palotti, J., Hanbury, A.: Query variations and their effect on comparing information retrieval systems. In: Proceedings of the 25th ACM International on Conference on Information and Knowledge Management, pp. 691–700. ACM (2016)

IR Evaluation

An Investigation into the Use of Document Scores for Optimisation over Rank-Biased Precision

Sunil Randeni[✉], Kenan M. Matawie, and Laurence A.F. Park

School of Computing, Engineering and Mathematics,
Western Sydney University, Sydney, Australia
{s.randeni,k.matawie,l.park}@westernsydney.edu.au

Abstract. When a Document Retrieval system receives a query, a Relevance model is used to provide a score to each document based on its relevance to the query. Relevance models have parameters that should be tuned to optimise the accuracy of the relevance model for the document set and expected queries, where the accuracy is computed using an Information Retrieval evaluation function. Unfortunately, evaluation functions contain a discontinuous mapping from the document scores to document ranks, making optimisation of relevance models difficult using gradient based optimisation methods. In this article, we identify that the evaluation function Rank-biased Precision (RBP) performs a conversion from document scores, to ranks, then to weights. Therefore, we investigate the utility of bypassing the conversion to ranks (converting document score directly to RBP weights) for Relevance model tuning purposes. We find that using transformed BM25 document scores in the place of the RBP weights provides an equivalent optimisation function for mean and median RBP. Therefore, we can use this document score based RBP as a surrogate for tuning relevance models.

Keywords: Information retrieval · Relevance model tuning · Rank-biased precision · BM25

1 Introduction

The task of an information retrieval system is to provide a set of candidate documents that are predicted to be relevant to a provided query. It is common for retrieval systems to provide a score to each document that reflects its relevance to the query, where the greater the score, the more likely the document is relevant. Rather than providing each document's score to the user, the documents are ordered by decreasing relevance score, allowing the user to first examine the most likely relevant document before proceeding to the rest, or reformulating the query.

Document scores are computed using a relevance model; a function that takes the query and a document, and returns a score reflecting the likelihood that the

© Springer International Publishing AG 2017
W.-K. Sung et al. (Eds.): AIRS 2017, LNCS 10648, pp. 197–209, 2017.
https://doi.org/10.1007/978-3-319-70145-5_15

document is relevant to the query. Many relevance models exist (e.g. BM25, TF-IDF, Language Models) and each has parameters that allow us to tune the model to the document set to optimise the retrieval effectiveness.

Tuning of mathematical model parameters is most efficient when we have access to the gradient of the optimisation function, allowing us to iterate towards the optimal point. But unfortunately, tuning retrieval model parameters usually requires a brute force approach, where we define the range of parameters, run a test for each combination of parameters, examine the accuracy of the retrieval results for each combination of parameters and then choose the combination of parameters that provide the best test results. This brute force approach is required due to the retrieval system evaluation function having a discontinuous gradient since it is based on the document rankings and not the document scores.

In this article, we investigate the potential for using document scores in the retrieval system evaluation function. Doing so will provide a smooth optimisation gradient, which in turn will allow more efficient optimisation and tuning. In particular, we examine the effect of using BM25 document scores as a surrogate for the Rank-biased precision (RBP) weight [13].

The article will proceed as follows: Sect. 2 describes the problems associated to tuning relevance models. Section 3 presents the related work. In Sect. 4, we examine how RBP can be modified to use document scores rather than ranks and discuss the experimental results with our analysis. Finally, Sect. 5 presents the conclusions and our directions for future work.

2   Tuning Relevance Models

The task of an Information Retrieval (IR) system is to supply a subset of documents that are relevant to a user supplied query, selected from a larger predefined document collection [4,11]. Rather than choosing an exact subset, it is common for retrieval systems to assign a score to each document denoting the document-query similarity. The retrieval system then proceeds to order the documents by their score in decreasing order (i.e., the most likely relevant at the top) and the documents are presented to the user accordingly.

An evaluation function is used to assess the accuracy of an Information Retrieval system. Evaluation functions require the document ranking produced by the retrieval system and set of manual relevance judgements. The evaluation function provides a score denoting the correctness of the IR systems ranking with respect to the manual relevance judgements.

Retrieval systems assign relevance scores to documents using a relevance model. The relevance model compares the user provided query to each document in the document collection to obtain a relevance score for each document. An example relevance model is BM25:

$$\mathcal{R}(d, Q; k_1, b) = \sum_{t \in Q} \left(\frac{f_{t,d}(k_1 + 1)}{f_{t,d} + k_1 \left(1 - b + b \frac{l_d}{l_{\text{avg}}}\right)} \right) w_t \tag{1}$$

where d is the document, Q is the set of query terms, $f_{t,d}$ is the count of term t in document d, l_d is the length (number of terms) in document d, l_{avg} is the average document length for the document collection, k_1 and b are model parameters, and w_t is the term weight, usually given as:

$$w_t = \log \left(\frac{N - df_t + 0.5}{df_t + 0.5} \right) \tag{2}$$

where N is the number of documents in the collection, and df_t is the number of documents that contain term t in the collection.

Each of the components of the relevance model is defined by the document collection, except for the model parameters. The relevance model parameters must be set by the administrator of the information retrieval system and are preferably set to optimise the accuracy of the relevance scores, using a training and testing query set. The relevance model provides a score to each document, but they are relative scores, since we must compare all document scores to identify which are the most likely relevant. Therefore, to optimise the relevance model, it must optimise per query, not per document. The optimisation is given as:

$$\operatorname*{argmax}_{\Theta} \mathcal{S}[\{e_1, e_2, \ldots, e_M\}]$$
$$\text{s.t. } e_i = \mathcal{E}(\boldsymbol{r}_i, \boldsymbol{s}_i)$$
$$s_{ij} = \mathcal{R}(d_j, Q_i; \Theta)$$

where $\mathcal{R}(d_j, Q_i; \Theta)$ is the relevance model with parameters Θ, s_{ij} is the relevance score for document j and query i, $\mathcal{E}(\boldsymbol{r}_i, \boldsymbol{s}_i)$ is the evaluation function comparing the computed vector of relevance scores $\boldsymbol{s}_i$ to the vector of relevance judgements $\boldsymbol{r}_i$, e_i is the evaluation score for query i and $\mathcal{S}[\{e_1, e_2, \ldots, e_M\}]$ is a statistic of the set of M evaluation scores. Therefore, we want the relevance model parameters Θ that maximise the chosen statistic of the set of evaluation scores. Note that a commonly chosen statistic is the mean of the query evaluations, but this may not be the best statistic to use to fit a model due to its sensitivity to outliers.

If we were to choose $\mathcal{S}$ to be the mean, or equivalently the sum, we then want to find Θ that maximises the sum of all e_i. If $\mathcal{R}$ and $\mathcal{E}$ are linear functions, the problem would collapse in to a linear equation. If instead we set $\mathcal{S}$ to be the median, we must find the Θ that provides minimum absolute deviation.

Optimisation problems can be efficiently solved using gradient based optimisation, as long as the gradient exists in the parameter search space, and the gradient is smooth. A major hurdle in performing this optimisation comes from the evaluation function $\mathcal{E}$ which converts the set of document scores $\boldsymbol{s}_i$ into a ranking, turning the set of numerical scores into ordinal values. This process removes the linearity from the optimisation problem, leaving us to resort to optimising over the set of all possible document ranking permutations [17].

To examine the problem in more detail, let us consider the Rank-biased Precision evaluation function (based on the RBP formula from [13]):

$$\mathcal{E}(\boldsymbol{r}, \boldsymbol{s}) = (1 - p) \sum_{i=1}^{N} r_i p^{\langle s_i \rangle - 1} \tag{3}$$

where $r_i \in \boldsymbol{r}$ is the ith manual relevance judgement, $s_i \in \boldsymbol{s}$ is the ith document score, $\langle s_i \rangle$ is the rank of document score s_i (the highest score is ranked first), N is the number of documents and $p \in [0, 1]$ is the function parameter. The operator $\langle \cdot \rangle$, mapping the numerical document scores into ordinal rankings, leads to the evaluation function not being smooth and so difficult to optimise.

Using the example in Table 1, there are five documents (with ID 1 to 5), each with a relevance score s_i (computed using a relevance model), manual relevance judgement r_i, the RBP weight for the given rank ($(1-p)p^{i-1}$ for rank i) and the RBP score at the bottom of the table. Note that the RBP weight only contributes to the RBP score if the associated relevance judgement is 1. An example of what might happen if we change the relevance model parameters is provided in Table 2. By changing the model parameters, we have provided better scores for each document. When compared to Table 1, the document scores in Table 2 have all moved in the correct direction (relevant documents have increased and irrelevant have decreased). But since the ranking of the documents did not change, the RBP score does not change. Since small changes in the parameters are likely to lead to no changes in RBP, gradient based optimisation methods will fail to help in optimising this problem.

Table 1. An example RBP computation on five documents. Each document score is converted into a rank, and then mapped to an RBP weight ($(1-p)p^{i-1}$, with $p = 0.8$). The RBP weight is then added to the final evaluation score if the document is relevant (Rel. Judg., r_i is 1).

Doc ID	Score (s_i)	Rel. Judg. (r_i)	RBP weight
1	20.1	0	~~0.20000~~
2	15.8	1	0.16000
3	12.6	1	0.12800
4	10.4	0	~~0.10240~~
5	6.8	1	0.08192
Total			0.36992

Table 3 shows what might happen if we adjust the relevance model parameters further. In this case, the parameter change has led to the score of document 2 increasing and the score of document 1 decreasing, leading to a change in rank the documents at rank 1 and rank 2. Unlike the previous example, where a change in relevance model parameters lead to no change in RBP, this change in parameters has led to a jump in RBP.

Table 2. Changing the relevance model parameters has pushed the document scores in the correct direction compared to Table 1, but the RBP evaluation score is unchanged.

Doc ID	Score (s_i)	Rel. Judg. (r_i)	RBP weight
1	19.7	0	~~0.20000~~
2	18.4	1	0.16000
3	17.8	1	0.12800
4	9.6	0	~~0.10240~~
5	8.2	1	0.08192
Total			0.36992

Table 3. Changing the relevance model parameters further has lead to the documents at rank 1 and 2 swapping (compared to Table 2), leading to a jump in the RBP score from 0.36992 to 0.40992.

Doc ID	Score (s_i)	Rel. Judg. (r_i)	RBP weight
1	18.4	0	~~0.16000~~
2	19.7	1	0.20000
3	17.8	1	0.12800
4	9.6	0	~~0.10240~~
5	8.2	1	0.08192
Total			0.40992

So unfortunately, the optimisation gradient is either zero (no change) or infinite (a jump), meaning that we are unable to use gradient based methods when optimising over RBP. This problem exists for all evaluation functions, since they all convert document scores into ranks.

Rather than converting the document score to a rank, then to a weight, we hypothesise that we can use the document score in the place of the RBP weight. We do not suggest that this will provide the same RBP score, but we will examine if doing so will allow us to perform an equivalent optimisation, providing us with an estimate of the optimal relevance model parameters Θ.

3   Related Work

There have been a considerable number of investigations on parameter tuning in IR relevance models [6,7,16,17]. We only discuss a few of them here. These investigations involve normalisation on components such as term frequency and document length in relevance models. In IR literature we cannot find any attempt to optimise free parameters such as b and k_1 in BM25 relevance model using score based approach. Almost all weighting models take the within-document term frequency tf, the number of occurrences of the given query term in the given document, into consideration as a basic factor for weighting documents.

He and Ounis [6] proposed a method for tuning of term frequency normalisation parameter(s), by measuring the normalisation effect on the within-document frequency of the query terms. The term frequency component of the BM25 formula has been used for their experiments with the parameter values of 0.75, 1.2 and 1000 for b, k_1 and k_3 respectively. The term frequency component tf_n of the BM25 formula is shown below:

$$tf_n = \frac{f_{t,d}}{k_1 \left((1 - b) + b(\frac{l_d}{l_{\text{avg}}}) \right)} \tag{4}$$

To illustrate the suitability of this new method, they applied it on Amati and Van Rijsbergens "normalisation method 2" [1–3]. Out of four normalisation methods proposed by Amati and Van Rijsbergen, the most robust and effective one is called "normalisation 2" or "second normalisation" [1]. The "normalisation 2" assumes that term frequency density is a decreasing function of the document length. In another experiment, He and Ounis examined a classical method of the term frequency tf normalisation tuning which is the *pivoted normalisation* approach proposed by Singhal et al. [15]. The idea of the *pivoted normalisation* is to fit the document length distribution to the length distribution of relevant documents and it requires relevance assessment on each given collection. He and Ounis [7] introduced it as a collection-dependence problem and suggested to re-calibrate the tf normalisation parameter for different collections and proposed a collection-independence measure, namely the normalisation effect, to indicate the optimal parameter settings on diverse collections. The proposed tuning method was evaluated on various TREC collections, for both the "normalisation 2" of the divergence from randomness (DFR) models and the BM25's normalisation method.

Many language model tuning techniques are centered around the issue of smoothing. The smoothing effect tends to be mixed with that of other heuristic techniques. The examples are "Dirichlet"and "Jelinek-Mercer"[9,10,12].

Taylor et al. [16] built on recent advances in alternative differentiable pairwise cost functions such as RankNet [5], and showed that these techniques can be successfully applied to tuning the parameters of an existing family of IR scoring functions such as BM25. They used NDCG (normalised discounted cumulative gain) [8] as their evaluation measure.

In the IR literature, although many investigations followed by new methods have been introduced for parameter tuning in IR relevance models, they were related to the areas such as term frequency, document length and smoothing. Yet they use the ranking list for effectiveness measures and no direct evidence can be found in relation to the utilisation of score in place of rank. Our work in this article appears to be the first study that proposes the use of score instead of ranking for optimisation purposes. Also, no evaluation measures (including RBP) have directly utilised document scores for relevance model tuning.

4   Ranking to Geometric Sequence (Score Based RBP)

Section 2 explained that the difficulty in tuning relevance models comes from the information retrieval evaluation function converting the numerical document scores to ordinal values. It also showed the RBP evaluation function and gave insight as to how we might be able to avoid the numerical to ordinal conversion.

Recall Eq. 3:

$$\mathcal{E}(\boldsymbol{r}, \boldsymbol{s}) = (1-p)\sum_{i=1}^{N} r_i p^{\langle s_i \rangle - 1}$$

$$= \sum_{i=1}^{N} r_i (1-p) p^{\langle s_i \rangle - 1}$$

$$= \sum_{i=1}^{N} r_i w_i$$

where we have each document contributing w_i to the final score if $r_i = 1$, or not contributing if $r_i = 0$.

If we consider only the document contribution w_i, we find that each document score s_i is mapped to its rank $\langle s_i \rangle$, which is then mapped to an element of the geometric sequence $(1-p)p^{\langle s_i \rangle - 1}$; numeric to ordinal to numeric. If we can bypass the ordinal conversion process and map the document scores straight to the geometric sequence, the tuning complexity will be reduced, leaving us with the *score based* evaluation function:

$$\mathcal{E}(\boldsymbol{r}, \boldsymbol{s}) = \sum_{i=1}^{N} r_i f(s_i) \tag{5}$$

where $f(s_i)$ is a transformation of the document score. In fact, we do not need equality; since we are using this function to tune the relevance function parameters, we only need this function to have its peak when using the same parameters that provides the RBP peak.

In this section, we will first examine the feasibility of using BM25 document scores in the place of the RBP weight. We will then proceed to identify the utility of using this score based form of RBP to tune the BM25 parameters.

4.1   Experimental Environment

Our experiments are conducted using the Associated Press document collection from TREC (Text Retrieval Conference) Disk 2, containing 79919 documents. Topics 51 to 200 and their associated relevance judgements from TREC 1, 2 and 3 were chosen as the experiment query set. Index construction and the BM25 relevance model are provided by Apache Lucene release 5.3.1 running on Linux (Oracle VM VirtualBox 5.1.12) server release 5.0. Lucene is configured to use the Standard Analyser (English dictionary), with the BM25 [12] relevance model. In addition, Lucene is configured to produce a ranked list for each query containing the top 1000 ranked documents. Code was also written to interact with Lucene libraries and perform our experiments. We will also be using RBP with $p = 0.8$, to represent the persistence of a Web search user [14,18].

4.2   Examining BM25 Document Scores

In this section, we want to assess how similar BM25 scores are to the RBP weights, so we can identify any transformations that are required to map one to the other.

The RBP weight is a geometrically decaying sequence, where the weight is based on the rank of a document:

$$w_i = (1 - p)p^{i-1}$$

If we take the log of the weight, we have a linear function of the rank i:

$$\log w_i = \log \left((1 - p)p^{i-1} \right)$$
$$= \log \left(\frac{1 - p}{p} \right) + i \log (p)$$

A plot of the linearity is shown in Fig. 1. Therefore, to assess the similarity of the form of each set of BM25 document scores and the RBP weights, we will measure the linearity of the ordered scores.

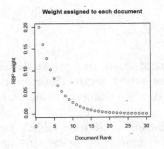
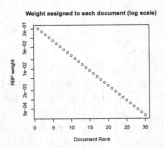

Fig. 1. The RBP weight for each rank is a geometrically decaying sequence, that is linear when using a log weight scale.

To examine the shape of the BM25 document scores, we will keep the BM25 parameters k_1 and b constant at 1.2 and 0.75 respectively. The sum of the set of RBP weights from rank 1 to ∞ is 1. Therefore, we normalise all sets of document scores to sum to 1. Also note that the RBP weight drops to 0.003 at rank 20, with the remainder of the weights above rank 20 contributing only 1.2% of the weight. Therefore, we will limit our analysis to the top 20 ranked documents.

To perform the analysis, we take the set of BM25 document scores for a given query, sort them in decreasing order of score, remove all document scores lower than rank 20, normalise the scores to sum to 1, take log of the scores, then fit a linear model. The linearity is assessed using the coefficient of determination ($R^2 \in [0, 1]$) of the linear model (a goodness of fit statistic). This process was performed on all 150 queries. An R^2 value of 1 implies that the log document scores are perfectly linear, while the lower the score, the less linear the document scores.

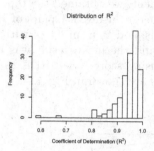

Fig. 2. Distribution of R^2 for the linear model log (document score) versus document rank, for the set of 150 queries.

A histogram of the 150 R^2 values is shown in Fig. 2. We find that the R^2 are mostly between 0.9 and 1, meaning that they are all highly linear. To gain a better understanding of the linearity the log document scores, we provide plots of the log document scores for the queries that have the minimum, first quartile, median, third quartile and maximum R^2 values as shown in Fig. 3.

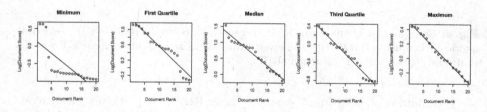

Fig. 3. The minimum, first quartile, median, third quartile and maximum linearity of the log document scores for the sample of 150 queries. The plots from left to right have R^2 values of 0.581, 0.927, 0.957, 0.974 and 0.992 respectively.

Each of the plots show highly linear log document scores, except for the plot showing the lowest R^2 score, where there is a large gap between the third and fourth documents. Since the majority of queries are highly linear, we will proceed to examining the effect of using these scores as weights when optimising the BM25 parameters.

4.3   Tuning BM25 Using Score Based RBP

The previous section showed that the BM25 document scores for the majority of queries had a similar form to the RBP weights when the top 20 documents are kept and the scores are normalised to sum to 1. In this section, we will compute the evaluation score for each query using both RBP and score based RBP (sRBP), over a range of BM25 parameters. This will allow us to view the optimisation surface and identify if both RBP and score based RBP provide maximal values at the same BM25 parameter sets.

To perform this experiment, we obtained the top 1000 documents and their BM25 scores for the 150 queries, using 341 pairs of BM25 parameters (using b from 0 to 1 with 0.1 intervals, and k_1 from 1 to 4 with 0.1 intervals). Each query was evaluated using RBP and the mean result was used to provide a contour plot. The same was done using score based RBP to obtain a contour plot for score based RBP. The plots are shown in Fig. 4.

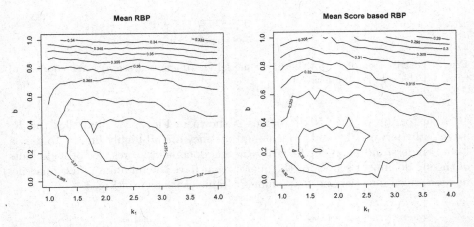

Fig. 4. Contour plot of the mean RBP and score based mean RBP (sRBP) over 150 queries, when using BM25 with the given values of k_1 and b. Left contour is for mean RBP and right contour is for mean sRBP.

The contour plots in Fig. 4 show the mean RBP (left plot) and mean score based RBP (right plot) obtained when using BM25 with the specified b and k_1. The plot of mean RBP shows that the optimal BM25 parameters are in the range $b \in [0.1, 0.4]$ and $k_1 \in [1.8, 3]$ giving the optimal RBP value of greater than 0.375. Using the score based RBP, we find that the optimal BM25 parameter range is $b \in [0.1, 0.3]$ and $k_1 \in [1.3, 2.5]$ giving the score based RBP value of greater than 0.33. The optimal regions are amazingly close for both RBP and score based RBP, implying that we can use score based RBP to optimise BM25 for mean RBP.

Figure 5 contains the contour plot for the median RBP (left) and median score based RBP (right). The optimal median RBP region is centred on $b = 0.6$ and

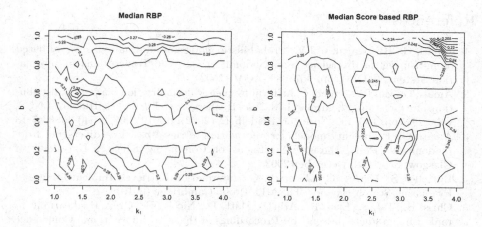

Fig. 5. Contour plot of the median RBP (left) and score based median RBP (sRBP, right) over 150 queries, when using BM25 with the given values of k_1 and b.

$k_1 = 1.4$. The optimal region for score based median RBP is $b = 0.6$, $k_1 = 1.7$. We find again that score based RBP(sRBP) has provided a good estimate of the optimal BM25 parameters.

These results demonstrate that score based RBP is a good candidate for optimising over RBP. The BM25 parameters chosen using score based RBP were not optimal for RBP, but were very close. Further investigation is required to determine if the estimates can be improved, and how optimal the results are in terms of unseen queries.

5   Conclusion and Future Work

Relevance models should be tuned to suit the document collection. Unfortunately, tuning by optimisation of an IR evaluation is difficult due to the evaluation function converting numerical document scores to ordinal ranks.

In this article, we observed that Rank-biased Precision takes a further step by converting the set of document ranks to weights. We investigated the effect of replacing the RBP weights with transformed document scores in place of these weights, avoiding the conversion to document ranks. By avoiding the conversion to ordinal values, we reduce the complexity of the relevance model tuning.

We found that the majority of queries provided document scores that closely followed the RBP weight pattern, allowing us to use document scores in the place of the RBP weights. Finally, we found that tuning the relevance model parameters using sRBP provided parameters that were close to those found using RBP. This provides evidence that the simpler to optimise score based RBP can be used as a surrogate for RBP when tuning relevance models.

Further work is required to determine if this property is valid for other relevance models and also out of sample queries.

References

1. Amati, G., Rijsbergen, C.J.V.: Probabilistic models of information retrieval based on measuring the divergence from randomness. In: ACM Transactions on Information Systems (TOIS), pp. 357–389. ACM (2002)
2. Amati, G., van Rijsbergen, C.J.: Term frequency normalization via pareto distributions. In: Crestani, F., Girolami, M., Van Rijsbergen, C.J. (eds.) ECIR 2002. LNCS, vol. 2291, pp. 183–192. Springer, Heidelberg (2002). doi:10.1007/3-540-45886-7_13
3. Amati, G.: Probability models for information retrieval based on divergence from randomness. Ph.D. thesis, Department of Computing Science, University of Glasgow, Scotland, December 2003
4. Büttcher, S., Clarke, C., Cormack, G.: Information Retrieval: Implementing and Evaluating Search Engines. The MIT Press, Cambridge (2010)
5. Chris, B., Tal, S., Erin, R., Ari, L., Matt, D., Nicole, H., Greg, H.: Learning to rank using gradient descent. In: Proceedings of the 22nd International Conference on Machine Learning, Bonn, Germany, pp. 89–96. ACM (2005)
6. He, B., Ounis, I.: A study of parameter tuning for term frequency normalization. In: CIKM 2003 Proceedings of the Twelfth International Conference on Information and Knowledge Management, New Orleans, Louisiana, USA, pp. 10–16. ACM (2003)
7. He, B., Ounis, I.: Term frequency normalisation tuning for BM25 and DFR models. In: Losada, D.E., Fernández-Luna, J.M. (eds.) ECIR 2005. LNCS, vol. 3408, pp. 200–214. Springer, Heidelberg (2005). doi:10.1007/978-3-540-31865-1_15
8. Järvelin, K., Kekäläinen, J.: IR evaluation methods for retrieving highly relevant documents. In: Proceedings of the 23rd Annual International ACM SIGIR Conference on Research and Development in Information Retrieval, Athens, Greece, pp. 41–48. ACM (2000)
9. Lafferty, J., Zhai, C.: Document language models, query models, and risk minimization for information retrieval. In: Proceedings of the 24th Annual International ACM SIGIR Conference on Research and Development in Information Retrieval, New Orleans, Louisiana, USA, pp. 111–119. ACM (2001)
10. Lafferty, J., Zhai, C.: A study of smoothing methods for language models applied to information retrieval. In: Proceedings of the 24th Annual International ACM SIGIR Conference on Research and Development in Information Retrieval, New Orleans, Louisiana, USA, pp. 334–342. ACM (2001)
11. Manning, D.C., Raghavan, P., Schütze, H.: Introduction to Information Retrieval, 2nd edn. Cambridge University Press, New York (2009)
12. Matawie, K., Hasso, S.: Information retrieval models: Performance, evaluation and comparisons for healthcare big data analytics. In: Proceedings of the 31st International Workshop on Statistical Modelling, Rennes, France, pp. 207–212 (2016)
13. Moffat, A., Zobel, J.: Rank-biased precision for measurement of retrieval effectiveness. ACM Trans. Inf. Syst. (TOIS) 27(1), 2:1–2:27 (2008)
14. Park, L.A.F., Zhang, Y.: On the distribution of user persistence for rank-biased precision. In: The Proceedings of the Twelfth Australasian Document Computing Symposium (2007)
15. Singhal, A., Buckley, C., Mitra, M.: Pivoted document length normalization. In: Proceedings of the 19th Annual International ACM SIGIR Conference on Research and Development in Information Retrieval, Zurich, Switzerland, pp. 21–29. ACM (1996)

16. Taylor, M., Zaragoza, H., Craswell, N., Burges, C.: Optimisation methods for ranking functions with multiple parameters. In: Proceedings of the 15th ACM international conference on Information and Knowledge Management, Arlington, Virginia, USA, pp. 585–593. ACM (2006)
17. Valizadegan, H., Jin, R., Zhang, R., Mao, J.: Learning to rank by optimizing NDCG measure. In: Advances in Neural Information Processing Systems, pp. 1883–1891 (2009)
18. Zhang, Y., Park, L.A.F., Moffat, A.: Click-based evidence for decaying weight distributions in search effectiveness metrics. J. Inf. Retrieval, 1–24 (2010)

Towards Privacy-Preserving Evaluation for Information Retrieval Models Over Industry Data Sets

Peilin Yang[1(⊠)], Mianwei Zhou[2], Yi Chang[3], Chengxiang Zhai[4], and Hui Fang[1]

[1] University of Delaware, Newark, USA
{franklyn,hfang}@udel.edu
[2] PlusAI, Los Altos, USA
mianwei@plus.ai
[3] Huawei Research America, Santa Clara, USA
yichang@acm.org
[4] University of Illinois at Urbana-Champaign, Champaign, USA
czhai@cs.uiuc.edu

Abstract. The development of Information Retrieval (IR) techniques heavily depends on empirical studies over real world data collections. Unfortunately, those real world data sets are often unavailable to researchers due to privacy concerns. In fact, the lack of publicly available industry data sets has become a serious bottleneck hindering IR research. To address this problem, we propose to bridge the gap between academic research and industry data sets through a privacy-preserving evaluation platform. The novelty of the platform lies in its "data-centric" mechanism, where the data sit on a secure server and IR algorithms to be evaluated would be uploaded to the server. The platform will run the codes of the algorithms and return the evaluation results. Preliminary experiments with retrieval models reveal interesting new observations and insights about state of the art retrieval models, demonstrating the value of an industry data set.

Keywords: Test collections · Privacy · Evaluation

1 Introduction

Evaluation is essential in the field of Information Retrieval (IR). Whenever a new IR technique is proposed and developed, it needs to be evaluated and analyzed using multiple representative data collections. Since the beginning of the field, there have been a few community-based efforts on constructing evaluation collections for the IR research, such as TREC, NTCIR and CLEF. These collections are available for researchers to download, and the researchers can then conduct experiments on these data collections using their own computers. Such an evaluation methodology has been used by many researchers in thousands of publications.

© Springer International Publishing AG 2017
W.-K. Sung et al. (Eds.): AIRS 2017, LNCS 10648, pp. 210–221, 2017.
https://doi.org/10.1007/978-3-319-70145-5_16

Although TREC collections can provide valuable insights on how well an IR method performs, they are not the same data collections used by the search engine industry. Unfortunately, privacy is one of the reasons that prevent industry from sharing their data sets [10]. As a result, it remains unclear how well the observations we draw about an IR method based on the TREC collections can be generalized to the real world data sets used in the search engine industry.

One possible solution is to anonymize the data to protect privacy [1]. However, the data anonymization would lead to the loss of some useful information, and it would also pose constraints on the developed methods. Recently, a data-centric evaluation methodology, i.e., privacy-preserving evaluation (PPE), has been proposed [3,5,13]. This evaluation methodology does not require the sharing of the industry data set, which protects the privacy of the data. Instead, it advocates the industry to host an online evaluation system so that the researchers could upload their codes to evaluate their effectiveness over the industry data sets. The proposed PPE framework is also related to the studies on Evaluation-as-a-Service (EaaS) [6–8], where users can leverage the APIs provided by the system to fetch documents or to submit a ranking request.

This paper follows the idea of the privacy-preserving evaluation (PPE) framework [5], and presents a specific implementation of the framework, i.e., the *PPE-M* system. With the implemented system, we evaluate a few representative basic IR models over an industry data set and compare the results with those obtained on the standard TREC collections. Our study demonstrates that the PPE framework enables researchers to evaluate their methods using industry data collections, which essentially closes the gap between the IR researchers in academia and the industry data. Moreover, evaluation over the industry data set makes it possible to gain new insights about existing retrieval models. We focus on the evaluation of basic IR models in this paper, but the framework can be easily generalized for other tasks.

2   A General Framework of Privacy-Preserving Evaluation

Traditional IR evaluation methodology often requires a data collection to be downloaded to a local computer that also stores the code of an IR algorithm. After downloading the data, we can then run the code, get the results on the data collection and conduct further analysis. Clearly, this methodology would not work well for industry data sets which are not publicly available.

To address this limitation, a data-centric based privacy- preserving evaluation framework has been recently proposed [5]. The basic idea is to keep a data collection securely stored on its own server while allowing researchers to upload their codes to the server. The codes can access the statistics of the data collection through some pre-defined strategies, and then will be executed on the host server. The results will be returned to the researchers for further analysis.

There could be many different ways to implement such a general framework. In particular, we propose 3 different levels of support for evaluation that can accommodate different trade-offs. The main idea is illustrated in Fig. 1. The top

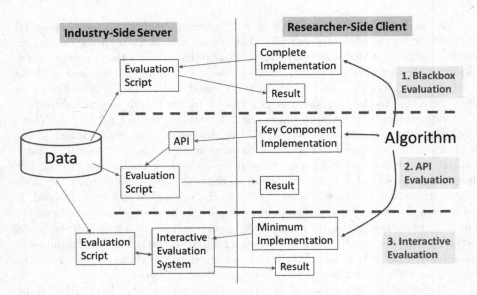

Fig. 1. Three-level support for PPE

level requires most work from the researcher's side but is most general as it can support evaluation of any algorithm in any language. Users of this kind of system have more access to the underlying system than other kinds of systems. For example, they could know how the index is built and thus are free to play with the index in order to better fulfill the ranking task. The middle level provides API support, so the researcher can focus on implementing just the key component of the algorithm to be evaluated, but it requires the researcher to use a particular API. The low level provides an interactive evaluation Web interface and attempts to minimize a researcher's work, but the algorithm that can be supported in this way may also be limited by the code that can be "opened up" by the system. An interactive system won't be able to provide full API support, so this would limit the algorithms that can be implemented and evaluated. It is clear that the top level is most general to support any algorithm, while the low level is most advanced with minimum effort on users, but has restriction on the algorithms to be evaluated.

We have already implemented the top level and the middle level, and will try to implement the low level in the future. Since the top level is trivial to implement, we focus on explaining how to implement the middle level in the next section.

3   A Specific Implementation

We now describe our implementation of the previously described privacy-preserving evaluation (PPE) framework. The implemented system focuses only on the evaluation of basic IR models, and is referred to as *PPE-M*.

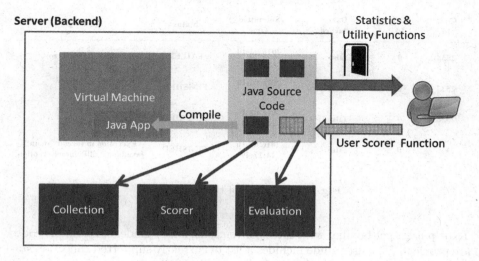

Fig. 2. System architecture

Source Code: [Choose File] No file chosen
Data File: [Choose File] No file chosen
[Task 1: Recency Evaluation ▼]
[Submit] [Clear]
Submission Status

Fig. 3. Screenshot of code submission interface

PPE-M is a web service with a typical client/server architecture. It hosts
data collections on the server, and enables users to implement and submit their
codes of retrieval models. Figure 2 shows the architecture of the implemented
system. Once a code is uploaded to the server, it will be executed to retrieve
documents from the collections. The retrieval results will be evaluated at the
back end, the evaluation results based on standard measures such as MAP will
be returned to the user for each data collection.

The front end of the system is a web form, which allows the users to upload
a Java source code file that includes the implementation of a retrieval function.
The screenshot of the code submission interface is shown in Fig. 3. Users can first
select which task to participate, and then upload the source code as well as the
data file (if necessary). The task could be ad hoc retrieval task, recency-based
retrieval task, etc. The source code includes the implementation of a retrieval
function in Java. The data file is optional, and it could include some prior information. Since the code will be executed on the server, it has to follow some
conventions on accessing the collection statistics and calling external functions.

Code Id ▾	Task Id	Data Attached	Submitted Time	Status	Message
2538	0	false	2016-02-10 14:27:32	FAILED	Compilation Error.
2537	0	false	2016-02-10 14:27:32	FINISHED	ndcg3:0.7145 map:0.8397
2536	0	false	2016-02-10 14:17:49	FINISHED	ndcg3:0.7145 map:0.8397
2535	0	false	2016-02-10 14:17:49	FINISHED	Exception in thread "main" java.lang.NullPointerException

Fig. 4. Screenshot of the result page

To help users get familiar with these conventions, a user guide and example codes are provided. The user guide includes a list of currently supported collection statistics that can be accessed by the code as well as a list of utility functions that the code can call. The restrictions posed on the codes are to prevent potential malicious attack from outsiders while making it possible for users to leverage the provided statistics to evaluate their models.

The core component of the *PPE-M* system is the Java source code package as the "Server (Backend)" rectangle. It consists of several modules, and each of them is responsible for a functionality. The *Collection* module basically preprocesses the collection and build the index of the collection. Key processes include tokenization, stop words removal, stemming, etc. It also provides the APIs to interact with the index so that the index does not necessarily need to be exposed to the users. The *Scorer* module is the base module for all ranking models. Models that are implemented and uploaded by the users have to complete some key functions that are required by the base *Scorer* module (think about this as derivation in object oriented programming). The *Scorer* module is then compiled and the binary is executed in the virtual machines. After the code execution, the *Evaluation* module kicks in. It computes the relevance scores of documents, generates the ranking list for each document and finally evaluate the model for different metrics. Such a modularized architecture offers flexibility in the implementation of the framework since each module could be re-implemented and tested independently.

The system returns the evaluation results to users through an interface as shown in Fig. 4. If the code can be compiled and executed correctly, the evaluation results will be returned, as shown in the last column. Otherwise, error messages will be displayed. Users are able to see how well their ranking models perform over each available data set. Moreover, they are able to see the evaluation results of codes submitted by others. In the future, we plan to further enhance the interface with a leader-board that sort and display all the submitted runs based on their performance.

The *PPE-M* system is implemented and designed in the above way for the following reasons. First, the system preserves the privacy of the industry data

Table 1. Statistics of test collections

Collections	IC	ROBUST04	WT2G	GOV2
#queries	3,274	250	50	150
avg(ql)	2.80	2.73	2.44	3.11
avg(idf(qt))	13.75	11.50	9.81	13.49
#documents	71,406	528,155	247,491	25,205,179
avg(dl)	583.44	467.55	1057.59	937.25

collections. The data collections are not distributed to users but are stored on the server. Users' code may only access certain type of information about the collection and use them to compute the relevance scores, but the collection information would not be passed to the client side. Second, the system is configurable based on the level of the privacy concerns about the data collections. For example, if more information can be released about a data collection, users can use the information in their codes or access more information from the evaluation results. Finally, the system can be easily generalized to evaluate other tasks in IR such as recency-based retrieval and click-prediction.

4 Experiments

We evaluate *PPE-M* using an industry data set, which contains 3,274 news-related queries and 71,406 articles. The queries were collected across a few months. For each query, around 20 documents are selected from all the news articles based on the ranking produced by a very simple retrieval method. For each query, editors manually assign each document with a relevance label (1-Bad, 2-Fair, 3-Good, 4-Excellent). In particular, we focus on using the industry data set to verify observations about basic retrieval models that people have made previously on TREC data sets. As the results will show, the new data set is useful since it can provide new insights on existing retrieval models.

4.1 Experiment Design

We denote the industry data set described earlier as *IC*. In addition to this data set, we also report results on a few representative TREC collections: *ROBUST04*, *WT2G* and *GOV2*. These data sets are selected to cover different types and sizes of the collections. The statistics of all the collections are summarized in Table 1.

We compare three representative retrieval functions: (1) Okapi BM25 (**BM25**) [9]: a function derived from the classic probabilistic model; (2) Pivoted document length normalization (**Piv**) [11]: a function derived from the vector space model; and (3) F2EXP (**F2EXP**) [4]: a function derived using axiomatic approaches. These three functions are selected because they are among the most effective retrieval functions based on the evaluation over multiple TREC collections. *F2EXP*, in particular, has been shown to be more robust than existing retrieval functions with comparable optimal performance. The main difference between *F2EXP* and other retrieval functions lies in its different implementation of IDF and document length normalization parts.

4.2 Retrieval Performance Comparison

We compare the optimal performance of the retrieval functions over all the data sets and summarize the results in Table 2. Figures 5, 6 and 7 show the parameter sensitivity curves.

In general, the results on the *IC* data set are consistent with those on the TREC collections. Specifically, the optimal performance of the three functions are comparable. The optimal parameters are also within the reasonable range as mentioned in the previous study [2]. However, there are also a few new interesting observations that we can make based on the results from the industry data set.

The first interesting observation is that the evaluation results on the *IC* data set are much higher than on the TREC collections. This is not surprising since

Table 2. Optimal performance comparison (MAP). Optimal parameter settings are reported in parenthesis.

Model	IC	ROBUST04	WT2G	GOV2
BM25	0.8687	0.2478	0.3152	0.2970
	(0.35)	(0.20)	(0.15)	(0.35)
PIV	0.8693	0.2206	0.2945	0.2536
	(0.20)	(0.05)	(0.05)	(0.05)
F2EXP	0.8595	0.2512	0.2973	0.2828
	(0.00)	(0.30)	(0.25)	(0.25)

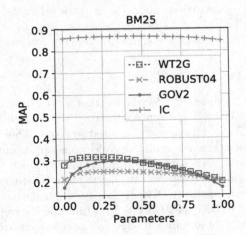

Fig. 5. Parameter sensitivity (MAP) for BM25

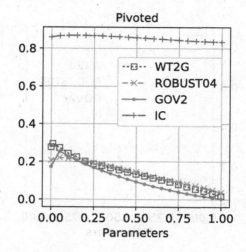

Fig. 6. Parameter sensitivity (MAP) for Pivoted

the *IC* data set is constructed by pooling top ranked documents for each query based on a simple ranking method while the documents of TREC collections are selected independently to the queries. Since most Web search engines now adapt a multi-level ranking strategy [12], the *IC* data set actually represents a more realistic problem setup.

The second interesting observation is about the *F2EXP* function. Although *F2EXP* has been shown to be robust in terms of the parameter values, its optimal parameter value is always larger than 0. However, its optimal parameter value is equal to 0 for the *IC* data set, which indicates that its length normalization part is not very effective. This is something that we have never observed based on the results for TREC collections.

4.3 Further Analysis

So far, we have demonstrated that the *PPE-M* system is able to evaluate retrieval models without releasing the data set. One new discovery made using this data set is about the "unusual" optimal parameter value in the *F2EXP* function.

To look into the reason behind this observation, we conduct more analysis using diagnostic evaluations [2]. The diagnostic evaluation methodology was proposed to identify the weaknesses and strengths of retrieval functions based on the perturbation of collections [2]. Each perturbation is designed to test a specific aspect of a retrieval function. Some perturbations can be done by changing simple statistics, while others may require additional information about the collections. In this paper, we only apply perturbation tests that can be implemented using the available statistics provided by the *PPE-M* system. These tests include two length variance sensitivity tests, one term noise resistance test and three TF-LN balance tests.

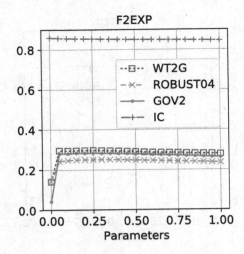

Fig. 7. Parameter sensitivity (MAP) for F2EXP

Figure 8 shows the perturbation results, split by the type of the perturbation. Here we choose three types of the perturbation, namely LV1, TN1 and TG3. For each type, we compare its results for *IC* data set with that of *GOV2* data set. We only show the results of the tests that are different on the two sets, so that we can focus on new insights gained by using the industry data set.

The first perturbation test is the length variance reduction test (LV1). We prefer curves that are lower because they indicate the functions would have more gain on length normalization part. Two plots on the first column in Fig. 8a indicate that *F2-EXP* has less gain on length normalization part for the *IC* data set, which is something we fail to observe from the TREC data set.

The second test is the term noise resistance test (TN). The curves that are higher means that they penalize long documents more appropriately. The plots on the second column in Fig. 8b suggest that *F2-EXP* did a poor job to penalize long documents with more noisy terms on the *IC* data set. However, such a trend is not clear based on the results from the TREC data set.

The third test is the all query term growth test (TG3). We prefer curves that are higher since it means the corresponding function can balance TF and LN more appropriately. The last column in Fig. 8c indicates that *F2-EXP* did a better job to avoid over-penalize long documents with more query terms on the *IC* data set. And this is something we can not see based on the results on the TREC data set.

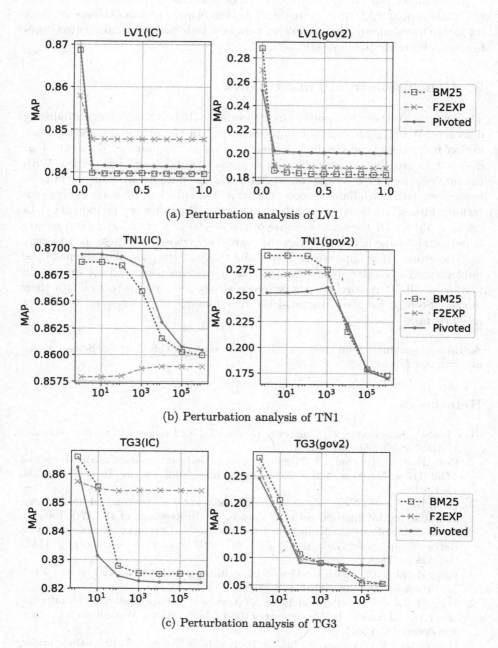

(a) Perturbation analysis of LV1

(b) Perturbation analysis of TN1

(c) Perturbation analysis of TG3

Fig. 8. Results of perturbation tests

In summary, our preliminary study has demonstrated the possibility of using the implemented *PPE-M* to evaluate IR models using a real world industry data set. More interestingly, we are able to gain new insights about existing retrieval functions by using the industry data.

5   Conclusions and Future Work

The paper addresses the issue of how to evaluate IR basic models with industry data sets. We present a specific implementation of the privacy-preserving evaluation framework, and conduct experiments on a real industry data set. This data set is usually not available for researchers outside the lab to use. With the implemented *PPE-M* system, we can make the data set available to outside researchers (still with limited access of course, but sufficient for some experimentation). Finally, we demonstrate that evaluating IR models on the industry data set is useful for IR research. Because of the availability of this data set, we are able to make some new discoveries that would otherwise be impossible to make.

There are a few interesting future directions. First, we plan to make the implemented system publicly available as a new open-sourced IR evaluation platform. All IR researchers are welcome to use the system to evaluate their models. Second, we plan to extend the functionality of this system to support more IR tasks.

Acknowledgments. This research was supported by the U.S. National Science Foundation under IIS-1423002.

References

1. Chor, B., Kushilevitz, E., Goldreich, O., Sudan, M.: Private information retrieval. J. ACM **45**(6), 965–981 (1998)
2. Fang, H., Tao, T., Zhai, C.: Diagnostic evaluation of information retrieval models. ACM Trans. Inf. Syst. **29**(2), 7–42 (2011). http://doi.acm.org/10.1145/1961209.1961210
3. Fang, H., Wu, H., Yang, P., Zhai, C.: Virlab: a web-based virtual lab for learning and studying information retrieval models. In: Proceedings of the 37th International ACM SIGIR Conference on Research & Development in Information Retrieval, pp. 1249–1250. SIGIR 2014, NY (2014). http://doi.acm.org/10.1145/2600428.2611178
4. Fang, H., Zhai, C.: An exploration of axiomatic approaches to information retrieval. In: Proceedings of the SIGIR 2005 (2005)
5. Fang, H., Zhai, C.: Virlab: a platform for privacy-preserving evaluation for information retrieval models. In: Proceeding of the 1st International Workshop on Privacy-Preserving IR (2014)
6. Hopfgartner, F., Hanbury, A., Müller, H., Kando, N., Mercer, S., Kalpathy-Cramer, J., Potthast, M., Gollub, T., Krithara, A., Lin, J., Balog, K., Eggel, I.: Report on the evaluation-as-a-service (eaas) expert workshop. SIGIR Forum **49**(1), 57–65 (2015). http://doi.acm.org/10.1145/2795403.2795416

7. Lin, J., Efron, M.: Evaluation as a service for information retrieval. SIGIR Forum **47**(2), 8–14 (2013). http://doi.acm.org/10.1145/2568388.2568390

8. Paik, J.H., Lin, J.: Retrievability in api-based "evaluation as a service". In: Proceedings of the 2016 ACM International Conference on the Theory of Information Retrieval, pp. 91–94. ICTIR 2016, NY (2016). http://doi.acm.org/10.1145/2970398.2970427

9. Robertson, S., Walker, S., Jones, S., Hancock-Beaulieu, M., Gatford, M.: Okapi at trec-3. In: Proceedings of TREC (1996)

10. Si, L., Yang, H.: Privacy-preserving ir: when information retrieval meets privacy and security. In: Proceedings of the SIGIR 2014 (2014)

11. Singhal, A., Buckley, C., Mitra, M.: Pivoted document length normalization. In: Proceedings of the SIGIR 1996 (1996)

12. Wang, L., Lin, J., Metzler, D.: Learning to efficiently rank. In: Proceedings of SIGIR 2010 (2010)

13. Yang, P., Fang, H.: A reproducibility study of information retrieval models. In: Proceedings of the 2016 ACM International Conference on the Theory of Information Retrieval, pp. 77–86. ICTIR 2016, NY (2016). http://doi.acm.org/10.1145/2970398.2970415

Evaluating Opinion Summarization in Ranking

Anil Kumar Singh[✉], Avijit Thawani, Anubhav Gupta,
and Rajesh Kumar Mundotiya

IIT (BHU), Varanasi, India
nlprnd@gmail.com

Abstract. We discuss the evaluation of rankings of documents that aim to summarize the overall opinion expressed in product reviews. Such a ranking can be used by e-commerce websites to represent a gist of the public opinion about a product. The inability of traditional IR metrics to reward such a representation is argued for. A matrix based labelling procedure serves as the framework for such evaluation. Three alternative metrics are adapted from previous similar works to evaluate opinion representativeness. Similarly, two metrics are adapted for exhaustiveness. Finally, we compare the robustness of these metrics.

Keywords: Opinion · Reviews · Diversity · Proportionality · Summarization

1 Introduction

Owing to the conformist tendencies of people, the domains of search result ranking and document summarization possess a great potential (and bear a great responsibility) in influencing popular opinion about a target entity. A popular application of such an opinion summary is an ordered list of product reviews on an e-commerce website (like the one shown in Table 1). This list may be perceived as a weighted summary of the opinions of people about the product. For example, if on searching for 'iPhone reviews', a user sees results (ordered by, say, date) that coincidentally happen to be against the product, then the user might form a perception of the general opinion around the world regarding that product. This perception may or may not be in line with the original composition of opinion worldwide.

What, then, should be the basis of document ranking in information retrieval methods? If 80 out of 100 reviews hold a specific opinion (e.g.: 'iPhone battery is long lasting'), should this be primarily reflected in the top-k reviews displayed on the front page? Should the reviewer's authority be taken into account? Should the overall opinion expressed in all the reviews of that product be summarized in the resulting list? The inherent ambiguity and discord in the way opinions are conveyed adds to the complexity of this task.

At present, not enough attention is given to opinion disambiguation even in search engines. Demartini [3] notes how it may happen that rankings based on

© Springer International Publishing AG 2017
W.-K. Sung et al. (Eds.): AIRS 2017, LNCS 10648, pp. 222–234, 2017.
https://doi.org/10.1007/978-3-319-70145-5_17

Table 1. A sample ranking of product reviews

Sterling Silver Cubic Zirconia Eternity Ring	
Product Reviews	**Rating**
1. Alex Date : 24/07/2016	
This ring is pretty, it can go good with another ring. It narrow and the stone size is small by it self. It would be a good thumb ring. Again nice ring that does not have alot of bling.	3.0/5.0
2. Bran Date : 20/07/2016	
The ring was a gift and my daughter loved it!!! It is very sparkly and fit just right! I would highly recommend this product.	4.0/5.0
3. Chau Date : 14/07/2016	
This ring is amazing for the price. It doesn't turn my finger white, and the sizing is great. It's just a little bling that isn't too flashy. I wear it as a thumb ring. I think it's really pretty, and very sparkly.	3.0/5.0
4. Dany Date : 29/06/2016	
This sterling ring is not too wide, it has a nice touch with the CZ all the way around, making it easier to wear for my wife, because she doesn't worry about it spinning and cutting into the fingers to the side. The CZ stones are recessed a bit making it pretty smooth.	1.0/5.0

the popularity of web pages (e.g., PageRank), on topical relevance, and even topic diversity get biased towards a certain opinion. It must be noted that a similar disambiguation of named entities is much more common, and appears even in commercial web search engines and online encyclopedias. For example, a Google search for the word 'Titanic' disambiguates results as 'Titanic the ship' and 'Titanic the movie'.

Most of the commercial websites currently incorporate ranking of product reviews on the basis of parameters like date, rating, etc. But none of these methods are sure to give a gist of all the relevant opinions expressed in the review-set of the product. Considering the large number of reviews posted about any product, it is usually not possible for a user to read all of them. The end user is more likely to be well-informed if all the relevant opinions mentioned in all the reviews of the browsed product are expressed in the first, say, 5 or 10 reviews of the product.

The aim of this paper is to settle upon a framework of evaluating opinion summarization. This includes both the annotation procedure required, and the metrics of ranking evaluation. We contribute by formalizing a tabular representation of opinion annotation, which forms a base for other metrics.

The scope of this paper is limited to summarization only in terms of proportionality (representativeness) and coverage (exhaustiveness). Also, we will test the said evaluation measures on product review rankings. Product reviews serve as a suitable ground for experimenting with opinion diversification. They are concise, targeted, opinionated (though sometimes descriptive and sometimes objective), diverse (in terms of the category of product), readily available as datasets, and easily comprehended (which makes annotating such data easier).

2 Dataset

The Aspect Based Sentiment Analysis task at SemEval 2014–2016 [8] was an initiative towards the objective evaluation of sentiment expressed in product reviews. In a wide enough corpora of 39 datasets, the task was manifold (for example, see first sentence of Review 3, Table 1):

1. Identification of entity E and attribute A for which the opinion was expressed. E and A should be chosen from predefined inventories (Entity: 'Sterling Silver Cubic Zirconia Eternity Ring'; Attribute: 'price' in this example).
2. Extraction of the linguistic expression which contains the opinion ('amazing for the price' in this example).
3. Assign polarity to EA pair: Positive, negative or neutral ('+' in the example).

   The above three subtasks were evaluated on both per-sentence and per-review level. A motivation behind such an evaluation could be the objectivity of the method. Assessor faults and human annotation subjectivity were targeted to be sufficiently low by providing comprehensive of guidelines to the annotators[1].

   The purpose of this paper, however, is to assess an evaluation strategy that optimizes the final ranking of documents, and not the intermediate (yet admittedly a crucial) step of opinion mining and sentiment analysis. Such a framework must reward systems that ultimately produce an opinion diversified ranking, regardless of the opinions mined. Avoiding judgment on the basis of exactly specified opinions thus has two targeted benefits:

1. Due to the absence of an inventory of aspects or opinions for the participants to 'identify', the systems must approximate 'new' aspects that vary enormously for different products. Thus, vague aspects in the form of topics modelled will be rewarded equivalently to another approach that, say, manages to match exact words to the subtopics retrieved.
2. A lack of evaluation in the intermediary steps will avoid the over-fitting of methodologies to replicate human annotated opinions which are, inevitably, subjective and variable. A ranking which misses certain opinions or identifies a few extra ones, may produce similar rankings, thus scoring equally well.

   The opinion labelled dataset which we use (part of the NLTK corpus as well) is derived from the 'Mining and Summarizing Customer Reviews' paper by Bing et al. [5]. It is a compilation of the first 100 reviews of 5 products from Amazon.com and C—net.com (9 and then 3 more were added in subsequent years [4,7]). These reviews were sentence-wise annotated with the following:

1. Feature about which the opinion is expressed, if any
2. Orientation of opinion (+ or −)
3. Opinion strength (on a scale of 1 to 3)

---

[1] http://alt.qcri.org/semeval2016/task5/data/uploads/absa2016annotationguidelines.pdf.

An example of annotation by the human taggers (the authors of the paper themselves) for a digital camera is:

"**affordability**[+3] ## while, there are flaws with the machine, the xtra gets five stars because of its affordability."

This is a fairly sound annotation procedure, barring the inherent flaws associated with subjectivity of human annotation. Now we formally define some terms frequently used throughout the rest of the paper:

1. **Corpus:** the set of reviews accessible for a particular product. (e.g. iPhone corpus may consist of 20 reviews)
2. **Opinion:** any perspective towards the product. (E.g. 'iPhone battery is long lasting')
3. **Overall Opinion Vector or Overall Vector:** A vector of size $n =$ number of different opinions present in the corpus, signifying the distribution of opinions present in all the reviews for a product. (e.g. we may have 20 reviews such that 10 of them suggest 'iPhone battery is great', while 10 of them suggest 'iPhone is expensive' and 5 of them suggest 'iPhone screen is bright'. Then $n = 3$ and the overall opinion vector is [10, 10, 5]. The vector may even be normalized or expressed in other ways. Note that each review may have more than one opinion in this case.)
4. **Ranking Opinion Vector or Opinion Vector (of a given ranking):** A vector of size n ($n =$ number of different opinions present in the corpus) signifying the distribution of opinions present in the ranking. (e.g. [3, 0, 0] is an opinion vector representation of a ranking that includes only 3 reviews, all of which only suggest 'iPhone battery is great')

For ease of evaluation, we introduce the opinion list (Table 2) and the **opinion matrix** (Table 3) representations used for the labelling procedure.

The opinion-document matrix (or simply the opinion matrix) is a tabular output of the labelling process. Such a matrix can be used to compute the overall and opinion vectors easily (see definitions above). Overall vector will be the cumulative projection of the matrix vector onto the horizontal axis (as shown in Table 3). Opinion vector of a ranking can be computed by following similar procedure (discounting with increasing rank) for only some of the rows in the opinion matrix.

3   Ranking Metrics

Out of the several kinds of ranking evaluation metrics used today, we would focus on only those that deal with subtopics instead of a single relevance score for each document.

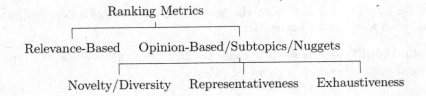

Table 2. Opinion list for "Sterling Silver Cubic Zirconia Eternity Ring" table

Opinion list
Realistic look
Good deal
Thumb ring replacement
Many sizes
Good for gifts
Matches with jewelry collection
Dainty and sparkly ring
Comfortable fit
Quality product
Long lasting and durable
Stones fallen out
Not much sparkly

Table 3. Opinion matrix for "Sterling Silver Cubic Zirconia Eternity Ring" table

Opinion matrix					
	Realistic look	Good deal	Many sizes	Good for gifts	Sparkly ring
Review1	X	X			X
Review2		X			
Review3			X	X	
Review4	X				X
Review5	X	X	X		
Review6	X			X	
Review7				X	X
Review8	X		X		
Review9					X
Review10		X	X	X	
Overall	5	4	4	4	4

3.1 Representative/Proportional

The following three metrics reward rankings that express opinions in the same proportion as that of the overall corpus.

(A) Distance Based Measures: The Characteristic Review Subset selection problem mentioned in Lappas 2012 [6] attempts to minimize the distance between a candidate opinion vector (vector V of set S) and a target vector. For a summarization cause, the candidate vector of set S is defined as:

$$V_S[i] = \frac{count(R|R \in S|n_i \in R)}{count(R|R \in S)}$$

The target vector will be the overall vector (O) of the corpus (S'). Lappas et al. maintain $L2$ norm as the only distance metric to judge the closeness of the two vectors. This can be generalized to produce the following formulation for the characteristic subset S^*

$$S^* = argmin_{S \in S'} D(V_S, O)$$

where $D(a, b)$ measures the distance between a and b, or

$$S^* = argmax_{S \in S'} C(V_S, O)$$

where $C(a, b)$ measures the closeness between a and b.

As an alternative to Euclidean distance being used as the distance metric (preceded by normalization of the two vectors in form of fractions instead of direct use of the count vectors of opinions in the sets), cosine similarity may be used in n-dimensional vector space (where n = number of opinions in the corpus for a particular product) as the closeness measure.

A major difference from our purpose is reflected in this approach, though. The CRS problem is a subset extraction problem and not a ranking problem. To adjust for a ranking evaluation measure, a discounted way of defining the opinion vector may be designed, in line with the Discounted Cumulative Gain. The cumulative opinion vector after rank k, is defined by:

$$CO(k, i) = \left\{ \begin{array}{l} CO(k-1, i) + 1/\log_2(k+1) \text{ if } o_i \in d_k \\ \\ CO(k-1, i) \text{ if } o_i \notin d_k \end{array} \right\}$$

where i denotes the ith position or opinion in the vector.

(B) Cumulative Proportionality: Cumulative Proportionality, an evaluation based on the proportional seat allocation problem in elections, was proposed by Dang and Croft [2]. It starts with a modified version of the Least Square Index (LSq) to avoid over-penalizing the over-representation of certain aspects and penalizing the occurrence of non-relevant documents. We will be using the

term opinion instead of aspect in this paper. Disproportionality at rank K, is thus defined as:

$$DP@K = \sum_{i=1}^{N} c_i(O[i] - V[i])^2 + \frac{1}{2}n_{NR}^2$$

where N is the number of opinions (number of columns in opinion matrix M or the size of opinion list L); $O[i]$ is the number of relevant documents that $opinion_i$ should have (O is the overall vector), and $V[i]$ is the number of relevant documents actually found for this aspect (V is the opinion vector for this ranking).

$$c_i = \left\{ \begin{array}{l} 1 \; O[i] \geq V[i] \\ 0 \; \text{otherwise} \end{array} \right\}$$

A normalized Proportionality score is also suggested:

$$PR@K = 1 - \frac{DP@K}{IdealDP@K}$$

where Ideal Disproportionality (i.e., maximum disproportionality) is given by:

$$IdealDP@K = \sum_{opinion_i} O[i]^2 + \frac{1}{2}K^2$$

Finally, Cumulative Proportionality (CPR) for evaluation of a ranked list on the basis of its proportionality is calculated as follows:

$$CPR@K = \frac{1}{k} \sum_{i=1}^{K} PR@i$$

(C) Weighted Relevance: For this last measure, we begin from the work of Clarke et al. [1], where a nugget-based framework for ranking evaluation is used and the formulation of a novelty-rewarding measure α -$nDCG$ is computed. Starting from an aspect-based interpretation of the Probability Ranking Principle, they proceed to deduce an expression where the user query drives the optimization of the measure. At this point, we depart from their process and instead, apply a summarization-based user query:

The fundamental goal of ranking is to predict the relevance of a document. To account for user interest, the probability that a document is relevant, is denoted by:

$$P(R = 1|u, d) = P(\exists o_i \text{ such that } o_i \in u \cap d)$$

where $o_i = i^{th}$ opinion, u = user query, and d = document.

The above expression assumes that the relevance of document d with respect to query u, is 1 if even a single aspect demanded by u, is exhibited by d. Next, assuming an independence among the occurrence of different aspects:

$$P(R = 1|u, d) = 1 - \Pi_{i=1}^{m}(1 - P(o_i \in u).P(o_i \in d)) \tag{1}$$

Now, we must determine $P(o_i \in d)$ and $P(o_i \in u)$. The probability of finding $opinion_i$ in document d is assessed by the framework in the previous section.

Assuming an assessor sub-optimal accuracy rate of α in predicting the presence of an opinion:

$$P(o_i \in d) = \left\{ \begin{array}{c} \alpha \text{ if } M(d,i) = \text{`}X\text{'} \\ 0 \text{ otherwise} \end{array} \right\}$$

where $M(d,i)$ is the N x D opinion matrix (N = number of opinions for the product, D = number of documents annotated, $1 \leq d \leq D, 1 \leq i \leq N$).

Thus, Eq. (1) is detailed into:

$$P(R = 1|u, d) = 1 - \Pi_{i=1}^{N}(1 - P(o_i \in u).\alpha M(d,i))$$

Lastly, we must reach a suitable definition of the probability of finding $opinion_i$ in the user query. Since, for the purpose of this framework, the user query is aimed at summarizing the opinions of the review corpus, the following definition is a reasonable one:

$$P(o_i \in u) = \frac{c(d|o_i \in d)}{c(d)} = \frac{O[i]}{D}$$

where $ç$ is the count function, O is the overall vector, and D is the number of documents (reviews). The fraction of documents in the corpus that possess opinion o_i are used to signify the probability of finding a relevant document which contains this opinion. For this reason, we call this measure the weighted relevance method of evaluation. The implication of the name gets clearer on further simplification of the equation:

$$P(R = 1|u, d) = 1 - \Pi_{i=1}^{m}(1 - \frac{c(d|o_i \in d)}{c(d)}.\alpha M(d,i))$$

$$= 1 - 1 + \frac{c(d|o_i \in d)}{c(d)}.\alpha M(d,i) + ... \approx \sum_{i=1}^{N} \frac{c(d|o_i \in d)}{c(d)}.\alpha M(d,i)$$

Dropping the constants α and $c(d)$ from the equation ($c(d) = D$, the total number of documents in the corpus) will leave the relative measure unaffected.

Thus, the jth element of the gain vector G (continuing from Clarke et al.) is defined as:

$$G[j] = \sum_{i=1}^{m} c(d_j|o_i \in d_j).M(d_j, i)$$

A discounted cumulative gain vector is then needed, to mould our relevance judgment into a full ranking evaluation:

$$DCG[k] = \sum_{j=1}^{k} \frac{G[j]}{log_2(1 + j)}$$

3.2   Exhaustive/Comprehensive

The following two metrics reward rankings that give equal representation to all major opinions of the corpus:

(A) S-Precision@R: As defined in Zhai et al. [9], S-Recall or subtopic recall at rank k is the fraction of subtopics retrieved after rank k. A subtopic i is said to be retrieved if any of the documents contains subtopic i. S-Recall for opinions in a document will be:

$$S - Recall@k = \frac{1}{N} \sum_{i=1}^{N} I(o_i \in D_1, D_2, ... D_k)$$

where N = number of opinions.

Digging deeper into traditional indexing algorithms, Ben et al. proceed to then adapting the S-Precision to aspect based ranking. S-Precision at Recall R is the minimum rank at which the S-Recall for the given ranking exceeds value R:

$$S - Precision@R = \min k$$
$$\text{such that } 1 \geq k \geq N \text{ and } S - Recall@k \geq R$$

(B) UnWeighted Relevance: For this measure, we retrace our steps to Eq. (1) in the previous subsection. For the modified problem of now calculating the coverage of a ranked list, we shall make necessary updates to the user query function.

$p(o_i \in u)$ must reflect the user's demand for opinion o_i to make it into the document d. Assuming equally important opinions, for the purpose of listing all possible opinions:

$$p(o_i \in u) = 1 \qquad \forall 1 \leq i \leq N$$

Thus the relevance of document d with respect to the coverage query u_c:

$$P(R = 1 | u_c, d) = 1 - \Pi_{i=1}^{m}(1 - \alpha M(d, i))$$

Again, reducing the equation:

$$P(R = 1 | u_c, d) = 1 - \Pi_{i=1}^{m}(1 - \alpha M(d, i))$$
$$= 1 - 1 + \alpha M(d, i) + ...) \approx \sum_{i=1}^{N} \alpha M(d, i)$$

This reduced form of relevance measure is interpreted as the number of opinions in document d, times the assessor accuracy α. Next, we define the gain vector and discounted gain vector same as above:

$$G[k] = \sum_{i=1}^{m} M(d_j, i)$$

$$DCG[k] = \sum_{j=1}^{k} \frac{G[k]}{log_2(1 + j)}$$

Despite having a different origin, the UnWeighted Relevance turns out to be a discounted form of S-Recall, if we make a minor change of not counting repeated instances of an opinion in the ranked list, and one that accommodates error in human annotation.

4   Experiments and Results

We have already listed out several evaluation measures in Sect. 3. Here, we will try to go deeper to understand which ones are better candidates for the evaluation of opinion representation and coverage.

Table 4. Correlation Matrix: comparison among 10 metrics over 10,000 random rankings each for 12 products (i.e. 120,000 samples in total)

Correlation Matrix									
	Euc	Cos	CPR	Wt	S-Pre	UnWt	ERR-IA	DCG-IA	RBP-IA	α-DCG
α-DCG	0.14	0.21	0.08	0.24	0.39	0.95	0.95	0.96	0.85	13.48
RBP-IA	0.19	0.07	0.01	0.36	0.23	0.88	0.96	0.83	0.52	
DCG-IA	0.32	0.13	0.07	0.40	0.38	0.98	0.91	21.29		
ERR-IA	0.17	0.13	0.04	0.30	0.32	0.94	2.21			
UnWt	0.29	0.12	0.02	0.41	0.35	458				
S-Pre	0.03	0.13	0.03	0.04	3.60					
Wt	0.82	0.56	0.02	37315						
CPR	0.09	0.03	308945							
Cos	0.75	0.01								
Euc	1.27									

4.1   EXPERIMENT: Correlation

In this first experiment, we try to find similarities between the available set of metrics. For any randomly selected ranking (out of all possible permutations), we calculate all metrics (α-DCG, ERR-IA, MAP-IA, RBP-IA, Distance-Euclidean, Distance-Cosine, CPR, Weighted Relevance, S-Precision, UnWeighted Relevance) and plot the correlations among these scores with each other. Total number of such samples taken were 10,000 random rankings for each product and there were 12 such products. The coefficients of correlation between each pair of metrics have been tabulated in Table 4 (e.g. CPR has a coefficient of correlation of 0.03 with Cosine similarity). Diagonal elements of the matrix represent the variance of the particular metric.

For a better analysis, let us divide the metrics into three classes: Novel/Diverse (α-DCG, ERR-IA, RBP-IA, DCG-IA); Representative (Euclidean Distance, Cosine Distance, CPR, Weighted Relevance); Exhaustive (S-Precision, UnWeighted Relevance).

Cumulative Proportionality and Weighted Relevance show a small correlation with every other metric, which can be attributed to their high variance. Among the others, a general trend shows high intra-class correlations and lower inter-class correlations. For example, Intent Aware metrics show a near 1 correlation among themselves. Another noteworthy anomaly is that of UnWeighted Relevance, which seems closer to the Intent Aware metrics than to the exhaustiveness-rewarding ones. This can be attributed to the fact that intent aware metrics show higher values when the number of subtopics covered is large, and UnWeighted Relevance mimics Subtopic Recall [9] (which will also reward large coverage of subtopics).

4.2   EXPERIMENT: Dependency

For the purpose of comparison of our proposed evaluation metrics, we chose to test the most crucial feature in the outlined framework: the opinion matrix. Over a collection of more than 80,000 annotated reviews (80 products) in the form of 80 opinion matrices, we calculated the values of all the metrics surveyed in Sect. 3. Such experiments were repeated for 50 arbitrary rankings of reviews (list-size=5) per product, and their average values were recorded. Also noted were the number of reviews/product (rows), no. of opinions as annotated by our lab members (columns) and how sparsely or densely the opinions appear in the matrix (Sparsity). Correlations of the values of all metrics with these factors are shown in Table 5.

Table 5. Dependency results

Dependency chart	Row	Column	Sparsity
alpha-DCG	0.11	0.16	−0.13
RBP_IA	0.09	0.03	−0.11
DCG_IA	0.10	0.02	−0.12
ERR_IA	0.10	0.19	−0.11
Un-weighted Relevance	0.09	0.02	−0.11
S-Precision	−0.05	**−0.27**	0.08
Wt. Relevance	0.16	−0.21	−0.18
CPR	−0.06	−0.11	0.06
Cosine Similarity	−0.06	**−0.29**	0.05
Euclidean	**0.30**	−0.13	**−0.31**

The intuition behind such an experiment is that metrics adapted to the opinion summarization may suffer from bias towards a longer, wider or denser opinion matrix. S-Precision has a noteworthy correlation with the number of columns, which is expected for a constant target Recall value (0.3 in our experiments). Cosine similarity is influenced by the number of dimensions/opinions/columns, while Euclidean, another distance metric is affected by number of rows/reviews. Despite being an intuitively promising metric, Cosine Similarity must hence be carefully used. Another option is to fix the number of annotated opinions/columns for all opinion matrices.

5 Conclusion

We raised the topic of opinion diversification in document ranking evaluation, along with a discussion on the dataset and metrics required for its evaluation. In this aspect, this paper can be thought of as a preliminary lab meeting for the opinion diversification community. We have discussed progress made so far and abstracted the best possible lessons from the previous work. Although we stop short of providing one ideal metric, we did compare the correlations and performance of the adapted metrics on the opinion matrix framework of evaluation.

A natural suggestion implied by this paper is an effective execution of a shared task. We are currently hosting such a shared task on Review Opinion Diversification at an international conference, on a product review dataset spanning 100 products and more than 10000 reviews. Evaluation strategies outlined above could be used to assess system performances, and the task runs can then be observed for insight into the respective pros and cons of the different measures of evaluation. Besides, there exist other areas worth looking into:

1. Reaching a suitable trade-off between relevance and opinion summarization in rankings, to be applicable in practical usage.
2. Weighted Opinions: What makes certain perspectives more important than the others? Can weights be applied to opinions for a form of biased representation in the summary?
3. User Study: An underlying assumption in our argument states that if a ranking of reviews proportionally represents the opinions or comprehensively covers all major opinions (creating a gist of the reviews), such a ranking will be of value to the user and the community. Counter arguments to this are welcome, and user studies are needed to verify such hypotheses.

References

1. Clarke, C.L.A., Kolla, M., Cormack, G.V., Vechtomova, O., Ashkan, A., Büttcher, S., MacKinnon, I.: Novelty and diversity in information retrieval evaluation. In: Proceedings of the 31st Annual International ACM SIGIR Conference on Research and Development in Information Retrieval, pp. 659–666. ACM (2008)
2. Dang, V., Bruce Croft, W.: Diversity by proportionality: an election-based approach to search result diversification. In: Proceedings of the 35th International ACM SIGIR Conference on Research and Development in Information Retrieval, pp. 65–74. ACM (2012)
3. Demartini, G., Siersdorfer, S.: Dear search engine: what's your opinion about...?: sentiment analysis for semantic enrichment of web search results. In: Proceedings of the 3rd International Semantic Search Workshop, p. 4. ACM (2010)
4. Ding, X., Liu, B., Yu, P.S.: A holistic lexicon-based approach to opinion mining. In: Proceedings of the 2008 International Conference on Web Search and Data Mining, pp. 231–240. ACM (2008)

5. Hu, M., Liu, B.: Mining and summarizing customer reviews. In: Proceedings of the Tenth ACM SIGKDD International Conference on Knowledge Discovery and Data Mining, pp. 168–177. ACM (2004)
6. Lappas, T., Crovella, M., Terzi, E.: Selecting a characteristic set of reviews. In: Proceedings of the 18th ACM SIGKDD International Conference on Knowledge Discovery and Data Mining, pp. 832–840. ACM (2012)
7. Liu, Q., Gao, Z., Liu, B., Zhang, Y.: Automated rule selection for aspect extraction in opinion mining. In: IJCAI, pp. 1291–1297 (2015)
8. Pontiki, M., Galanis, D., Pavlopoulos, J., Papageorgiou, H., Androutsopoulos, I., Manandhar, S.: Semeval-2014 task 4: aspect based sentiment analysis. In: Proceedings of SemEval, pp. 27–35 (2014)
9. Zhai, C.X., Cohen, W.W., Lafferty, J.: Beyond independent relevance: methods and evaluation metrics for subtopic retrieval. In: Proceedings of the 26th Annual International ACM SIGIR Conference on Research and Development in Information Retrieval, pp. 10–17. ACM (2003)

Author Index

Printed in the United States
By Bookmasters